I0056840

The Boron Arsenides

David J. Fisher

Copyright © 2023 by the authors

Published by **Materials Research Forum LLC**
Millersville, PA 17551, USA

All rights reserved. No part of the contents of this book may be reproduced or transmitted in any form or by any means without the written permission of the publisher.

Published as part of the book series
Materials Research Foundations
Volume 138 (2023)
ISSN 2471-8890 (Print)
ISSN 2471-8904 (Online)

Print ISBN 978-1-64490-222-6
ePDF ISBN 978-1-64490-223-3

This book contains information obtained from authentic and highly regarded sources. Reasonable efforts have been made to publish reliable data and information, but the authors and publisher cannot assume responsibility for the validity of all materials or the consequences of their use. The authors and publishers have attempted to trace the copyright holders of all material reproduced in this publication and apologize to copyright holders if permission to publish in this form has not been obtained. If any copyright material has not been acknowledged, please write and let us know so we may rectify in any future reprint.

Distributed worldwide by

Materials Research Forum LLC
105 Springdale Lane
Millersville, PA 17551
USA
http://www.mrforum.com

Printed in the United States of America
10 9 8 7 6 5 4 3 2 1

Table of Contents

Introduction

It was long thought that the biggest menace to the continued development of silicon-based semiconductors was Moore's Law: the yearly doubling of the number of components on a chip would surely imply a corresponding shrinkage of the components until they finally contained too few atoms to construct anything useful. Moore's law no longer seems to apply, and chip sizes are heading towards 1nm. When quantum-computing is considered, Moore's Law (2^n) is replaced by Neven's Law (2^{2n}).

There is another, more physical, problem however and this is that the heat generated by the ever more densely packed chips has to be removed. This task is not rendered easy, due to the poor thermal conductivity of silicon (table 1). Excess heat damages the circuitry and reduces performance and life-span. As components are made smaller, their resistance increases and so heat dissipation is a very important factor in semiconductor technological progress. It would also be advantageous to have a new material which not only had a high thermal conductivity but also offered higher, and more equal, hole and electron mobilities.

Such a wonder material exists, in the form of cubic boron arsenide: a III-V compound semiconductor with an indirect band-gap and a face-centered cubic zincblende structure. It offers the desired high electron and hole mobilities, together with a remarkably high thermal conductivity; nearly 10 times higher than that of silicon. It has already been hailed as 'the best semiconductor material ever found'.

General rules were established many years ago[1] for predicting which non-metallic crystals should exhibit a high thermal conductivity. The rules were:

low average atomic mass

strong interatomic bonding

simple crystal structure

low anharmonicity

Density functional theory predictions were later used to evaluate the effects of atypical phonon dispersion. Such dispersion could have a significant effect on 3-phonon scattering-rates due to selection rules. A large frequency-gap between the acoustic and optical phonon branches, the a-o gap, reduced the available phase space for 3-phonon scattering processes which involved 2 acoustic phonons and 1 optical phonon. A large a-o gap exists in materials which are made up of heavy and light atoms. The bunching of acoustic phonon branches moreover increases the thermal conductivity. Acoustic phonon bunching occurs when longitudinal and transverse phonons have similar frequencies within an appreciable volume of the Brillouin zone. The similarity of frequencies causes weak 3-phonon scattering between longitudinal and transverse acoustic modes. A small phonon scattering phase space leads to a high thermal conductivity.

Boron arsenide is an unusual thermal conductor due to the importance of 4-phonon scattering processes. In most other crystals, 4-phonon scattering events have no observable effect upon thermal transport. Density functional theory predictions suggest that 3-phonon and 4-phonon scattering rates are comparable in BAs. The temperature dependence of the thermal conductivity of BAs also suggests that 4-phonon scattering processes are important. When the thermal conductivity is limited by 3-phonon scattering rates, the thermal conductivity is proportional to 1/T at high temperatures. Between 150 and 600K, BAs exhibits a temperature dependence which is more sharp than 1/T. This agreed with density functional calculations which included the effects of 4-phonon scattering.

The drawback is that suitably large and uniform crystals of the material prove to be hard to produce. But even before it can become a viable replacement for silicon, its high conductivity may make it useful in those situations where diamond is currently used. Materials of high thermal conductivity dissipate heat and improve the reliability of devices. Diamond, with its thermal conductivity of 2000W/mK, is used for cooling but is slow to synthesise with a uniform quality and is expensive. It is also difficult to match to silicon, because of the difference in their thermal expansions. Other obvious alternatives such as copper and silicon carbide cannot really compete, given their relatively low room-temperature thermal conductivities of 400 and 350W/mK, respectively.

When samples of BAs were grown[2] by using a modified chemical vapor transport method the thermal conductivity was 1000W/mK: that is, 3 times higher than that of silicon carbide, but still lower that of diamond and the basal plane of graphite. Such a high conductivity in BAs is surprising because a high thermal conductivity can theoretically occur only in compounds comprising strongly-bonded light elements, due to the limiting effect of anharmonic 3-phonon processes. The point is that arsenic is regarded as being a heavy element. It was noted many decades ago that, because heat conduction in materials which have no mobile electrons occurs via phonons and because a low phonon scattering ensures a low thermal resistance, the best materials would prove to be those having a simple crystal made of tightly-bonded light elements. In these tests[3], a local room-temperature thermal conductivity of above 1000W/mK, with an average bulk value of 900W/mK, was found for bulk BAs crystals. The high values were explained in terms of the operation of higher-order phonon processes.

Table 1. Coefficients of thermal expansion and
thermal conductivities of various materials

Material	a (Å)	CTE (10^{-6}/K)	κ (W/mK)
GaAs	5.652	5.73	55
Si	5.431	2.63	130
GaN (a-axis)	3.190	5.59	200
GaN (c-axis)	5.189	3.17	200
SiC	4.348	3.8	360
Cu	3.597	16.5	401
BP	4.540	3.2	490
BAs	4.776	3.6	1300
Diamond	3.567	0.8	2200

A revision of the theory suggested that a high thermal conductivity could nevertheless be exhibited by crystals which were composed of one heavy, and one light, atom. The mass-difference between boron and arsenic also has the advantage of creating an energy-gap between the optical and acoustic phonons. The optical phonons are associated with out-of-phase vibrations between neighbouring atoms. The acoustic phonons are associated with in-phase vibrations. The large energy-gap impedes their interaction, so that they can then both travel easily through the lattice. An early application of the revised theory, which took account of the scattering among up to 4 phonons, rather than 3 phonons, predicted that the thermal conductivity of BAs should approach 1300W/mK in perfect crystals. The difference is due to the fact that 3-phonon combinations dominate scattering in crystals with a small energy-gap but cannot occur across the large energy-gap of BAs. Inclusion of fourth phonon widens the number of possible combinations and this governs the scattering in BAs.

All of this depends of course upon the perfection of the host crystal, and the preparation of high-quality bulk boron arsenide crystals proves to be difficult. Defects and impurities markedly reduce the thermal conductivity by decreasing the energy-gap between the phonons. One promising synthesis route was chemical vapor transport, but the high volatility of As tended to lead to the introduction of defects and alternative atom combinations. The products then offered disappointing thermal conductivities of the order of 190W/mK, although levels of over 1000W/mK were eventually obtained at room temperature. Continued progress in crystal growth now permits room-temperature values of 1300W/mK to be almost routinely measured in high-quality mm-sized BAs single crystals which are prepared by chemical vapor transport[4]. The undoped BAs is p-type conducting due to the presence of intrinsic acceptor defects.

In an early study[5], samples which were produced by chemical vapour deposition exhibited various degrees of structural disorder; from complete disorder, with completely random occupation of the zincblende structure sites to complete order, with almost all of the atoms on the correct sites. The reactions used were:

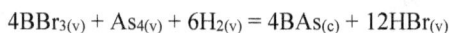

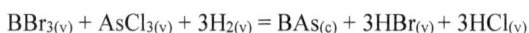

$$4BBr_{3(v)} + As_{4(v)} + 6H_{2(v)} = 4BAs_{(c)} + 12HBr_{(v)}$$

$$BBr_{3(v)} + AsCl_{3(v)} + 3H_{2(v)} = BAs_{(c)} + 3HBr_{(v)} + 3HCl_{(v)}$$

In the case of the second reaction, helium was used to carry the AsCl$_3$ to the reaction. Below 1000C, the phases which were produced were amorphous. At 1020C, the second reaction produced some crystalline BAs within a largely amorphous mass. Between 1035 and 1075C, crystalline BAs was produced as a minor phase within mixtures of amorphous material together with As or crystalline B$_{13}$As$_2$. Between 1100 and 1150C, crystalline B$_{13}$As$_2$ was the only product. Direct combination of the elements at close to atmospheric pressure, or at high pressures, also produced BAs.

All of the samples exhibited a very high degree of order, being assigned an order parameter of 0.90 to 0.95 and higher. A low-pressure method produced the most order (table 2).

Table 2. Relative peak intensities for BAs

Growth Method	Plane	Relative Peak Intensity*
Low-pressure	(200)	45.40
Low-pressure	(220)	36.51
Low-pressure	(222)	8.77
Low-pressure	(311)	39.89
Low-pressure	(331)	13.24
Low-pressure	(333)	13.59
Low-pressure	(400)	4.35
Low-pressure	(420)	10.63
Low-pressure	(422)	11.96
Low-pressure	(440)	8.01

Low-pressure	(442)	16.20
Low-pressure	(511)	13.59
Low-pressure	(531)	28.16
Low-pressure	(600)	16.20
High-pressure	(200)	29.18
High-pressure	(220)	31.70
High-pressure	(222)	13.37
High-pressure	(311)	38.20
High-pressure	(331)	13.57
High-pressure	(333)	19.65
High-pressure	(400)	9.26
High-pressure	(420)	9.53
High-pressure	(422)	14.85
High-pressure	(440)	9.54
High-pressure	(442)	21.06
High-pressure	(511)	19.65
High-pressure	(531)	44.71
High-pressure	(600)	21.06
Ideal order	(200)	39.60
Ideal order	(220)	39.54
Ideal order	(222)	9.45
Ideal order	(311)	39.64
Ideal order	(331)	16.77
Ideal order	(333)	18.21
Ideal order	(400)	6.22
Ideal order	(420)	13.04
Ideal order	(422)	16.23
Ideal order	(440)	9.79
Ideal order	(442)	26.03
Ideal order	(511)	18.21
Ideal order	(531)	44.10
Ideal order	(600)	26.03

$*I(hkl)/I(111) \times 100$

Although cubic arsenide has attracted a great deal of attention, other forms of the arsenide also exist, and these also exhibit intriguing properties.

History

This III-V boron compound remained unexplored for some time because of difficulties with the syntheses of pure single crystals. The first reported[6] preparation of boron arsenide was performed by directly reacting the elements in evacuated silica tubes for 12h at 800C. The compound had a cubic zincblende structure with a lattice constant of 4.777Å. (Table 3). It decomposed when heated in air but, in the presence of arsenic vapour, it was stable at up to 92 0C. At above that temperature it transformed irreversibly into a face-centered orthorhombic modification with a = 9.710, b = 4.343 and c = 3.066Å. This phase was very stable. The X-ray diagram did not noticeably change during lengthy boiling in alkali solution, concentrated HNO_3, HF or other acids.

Table 3. Powder diffraction data for cubic BAs

d(Å)	[hkl]
2.76	111
2.388	200
1.688	220
1.440	311
1.378	222
1.194	400
1.096	331
1.068	420
0.9752	422
0.9194	511/333
0.8445	440
0.8075	531
0.7962	442/600

Methods were found[7] for preparing compounds of the form, BAs and B_6As, as well-crystallized, thermally stable and chemically inert materials. It was also aimed to produce new chemically-resistant forms of boron arsenide as granular crystalline particles. In addition to any semiconducting properties, it was suggested that articles such as chemical apparatus and turbine blades might be made from boron arsenide. The various forms of boron arsenide were easily prepared by exposing elemental boron to arsenic vapour at high temperatures. amorphous boron is available commercially in the finely-divided state and may be directly. The reaction was somewhat slower for the crystalline form of boron. The reaction temperature between the boron and arsenic was generally above about 700C and could be as high as 1200C or higher. The time which was required for the reaction depended upon the temperature, upon the degree of sub-division of the boron and the degree to which the arsenic vapour could penetrate the mass of boron powder. The boron arsenides which were obtained had two distinct forms. The reaction of elemental boron with elemental arsenic at below about 1200C gave an orthorhombic crystalline form of material. Some variation in the stoichiometric formula was possible. It was found that the cubic crystalline form of BAs could be transformed by heating it at 900 to 1000C in order to obtain the orthorhombic crystalline product. The present BAs compound was stable against decomposition and was expected to melt at temperatures greater than about 2000C.

Both cubic and orthorhombic boron arsenide reparations were stable against reagents such as aqua regia. The orthorhombic material was also resistant to boiling nitric acid. It was suggested that articles such as chemical apparatus, crucibles and turbine blades, could be easily fabricated from orthorhombic boron arsenide.

Binary rocksalt and zincblende compounds exhibit lattice thermal conductivities which range over 3 orders-of-magnitude. Understanding of the fundamental heat-transfer mechanism is difficult because of the need to account for higher-order anharmonicity. Quartic anharmonicity plays an essential role in suppressing heat transfer in zincblende boron arsenide, with its potentially ultra-high conductivity. A then-novel high-throughput phonon framework, based upon first-principles density functional theory calculations was used[8] to investigate the room-temperature lattice dynamics and thermal transport properties of 37 binary compounds with rocksalt and zincblende structures. Particular attention was paid to the influence of quartic anharmonicity. A sophisticated theoretical model was used to predict the conductivity by incorporating state-of-the-art methods and included a complete treatment of quartic anharmonicity for phonon frequencies and lifetimes at finite temperatures, together with contributions arising from off-diagonal terms in the heat-flux operator. The effect of quartic anharmonicity upon conductivity was very different in the rocksalt and zincblende compounds, due to competing effects

Materials Research Forum LLC

https://doi.org/10.21741/9781644902233

upon the finite temperature-induced shifts in phonon frequencies and scattering rates. Correlation of the conductivity with the phonon-scattering phase-space provided a qualitative method for assessing the importance of 4-phonon scattering on the basis of harmonic phonon calculations. In the case of zincblende HgTe there was a sixfold reduction in the conductivity which was due to a 4-phonon scattering which outweighed 3-phonon scattering in the acoustic region at room temperature. There was also a potential breakdown of the phonon-gas model in rocksalt AgCl, where phonon states were appreciably broadened due to a strong intrinsic anharmonicity. This introduced off-diagonal contributions to the conductivity which were comparable to the diagonal contributions. Density functional theory calculations and Raman experiments revealed[9] the surprising fact that fourth-order anharmonicity generally plays an important or predominant role, as compared with that of third-order anharmonicity, at room temperature. This was even more the case at high temperatures for materials such as diamond, silicon, germanium, GaAs, boron arsenide, cubic silicon carbide and α-quartz. It was rendered possible by the large 4-phonon scattering phase-space of zone-center optical phonons. Raman linewidth measurements of BAs confirmed the predictions. The infra-red optical properties which were expected on the basis of the Lorentz oscillator model, after including 4-phonon scattering, were in much better agreement with experiment than were those which were based upon 3-phonon predictions.

Both the cubic and the rhombohedral forms of boron arsenide were later prepared by using vapour-phase techniques. Two microcrystalline forms of boron arsenide were again found. The lower-temperature product was cubic boron mono-arsenide, BAs, while the high-temperature material was rhombohedral boron sub-arsenide with lattice parameters of 4.7778 and 5.3177Å (α = 70° 32'). When the boron mono-arsenide was heated to above 1000C under arsenic pressures ranging from tenths of an atmosphere to 3atm the BAs transformed into the sub-arsenide. No transformation from rhombohedral to cubic occurred at 850C; even under arsenic pressures of up to 5atm.

The growth of BAs single crystals at the centimetre scale has long posed problems, thus greatly hindering research into its potential applications. One technique[10] permitted the growth of 7mm-long single crystals via chemical vapour transport, using heteronucleation sites. Differing thermal conductivity values which were measured for monocrystals that had been grown on various heteronucleation sites indicated the importance of choosing the optimum nucleation substrate.

Icosahedral Boron Arsenide

It was found[11] that cubic boron arsenide could be deposited onto silicon substrates by using a PH_3-B_2H_6-H_2 system. The 'lower' arsenide, $B_{13}As_2$, could be easily deposited by using an AsH_3-B_2H_6-H_2 system, with the orientation relationships,

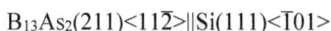

$$B_{13}As_2(110)<1\bar{1}0>\|Si(100)<100>$$

$$B_{13}As_2(110)<1\bar{1}0>\|Si(110)<1\bar{1}0>$$

$$B_{13}As_2(211)<11\bar{2}>\|Si(111)<\bar{1}01>$$

but BAs could not be deposited in that way. Given the early date of this work, it is unclear whether this phase had been accurately characterized.

With its band-gap of more than 3.0eV and its ability to self-heal radiation damage, icosahedral boron arsenide, $B_{12}As_2$, is an ideal candidate for use in future betavoltaic devices By converting high-energy electrons from radio-isotopes into electricity, devices made from $B_{12}As_2$ might power applications for decades. The arsenide, $B_{12}As_2$, is a member of a group of boride crystals having R3m symmetry. Classical valence-force-field analysis was used[12] to account for the observed Raman and infra-red spectra. The valence-force-field calculated modes for $B_{12}As_2$ were correlated with α-boron modes via eigenvector expansions, thus permitting the identification of modes that remained mainly B_{12} cluster modes and helped to clarify the effects of crystal forces and arsenic-atom interactions and movements. An attempt was made to account for the infra-red intensities by using the displacements of atoms. Qualitatively good agreement was found between the predicted and observed intensities. Dispersion curves were plotted from the Brillouin-zone center to the edge along two distinct symmetry axes. The acoustic-branch speeds ranged from 6 x 10^5 to 9 x 10^5cm/s for LA modes and were close to 4 x 10^5cm/s for TA modes. The compressibility of the boron sub-arsenide, $B_{12}As_2$, was studied[13] by using synchrotron X-ray diffraction at up to 47GPa at room temperature in a diamond anvil cell with neon as a pressure-transmitting medium. Fitting the experimental data by using the Vinet equation-of-state yielded a bulk modulus of 150GPa, with its first pressure-derivative being 6.4. No pressure-induced phase transitions were observed.

The phase transition was studied[14], using density-functional theory calculations, with regard to structural phase stability, elastic properties, sound velocity, Debye temperature and melting-point at zero and high pressures. The transition pressure from zincblende to NaCl phases was predicted, and was in good agreement with experimental data. At pressures below the transition pressure the zincblende phase was thermodynamically and mechanically more stable than was the NaCl phase. The mechanical behaviour was examined in terms of ductility and brittleness and differed only with regard to the exact boundary between the 2 types of mechanical behaviour. The variation of the longitudinal sound velocity as a function of pressure indicated that there was a softening of the corresponding phonons.

The thermal conductivity of boron arsenide is influenced by phonon scattering selection rules due to its special phonon dispersion. Compression leads to marked changes in the phonon dispersion, and is a testbed for first-principles theories of how phonon dispersion affects the 3-phonon and 4-phonon scattering rates. The pressure dependence of the thermal conductivity of BAs offers the opportunity to explore experimentally the relationship between acoustic bunching, 3-phonon and 4-phonon scattering, and thermal transport. Density functional theory calculations predict that the frequencies of longitudinal and transverse acoustic phonon branches separate under increasing pressure. That is, pressurization of BAs reduces acoustic bunching and makes its vibrational properties more like those of a typical crystal. Density functional theory predicts that 3-phonon scattering rates will increase with pressure, leading to a rapidly decreasing thermal conductivity. Density functional theory calculations also predict however that 4-phonon scattering rates decrease during pressurization. When both 3-phonon and 4-phonon scattering rates are included, density functional theory predicts that total phonon-phonon scattering rates depend weakly upon pressure when it is less than 20GPa. When the pressure becomes greater than 20GPa, it is predicted that the thermal conductivity will gradually decrease. The thermal conductivity was measured between 0 and 30GPa by using time-domain thermoreflectance methods and a diamond anvil cell. Unlike non-metallic crystals, the BAs exhibited a pressure-independent thermal conductivity below 30GPa. The thermal conductivity of non-metallic crystals instead increases upon compression. The anomalous pressure-independence of the BAs thermal conductivity reflects the important relationship between phonon dispersion and 3-phonon and 4-phonon scattering-rates.

New high-pressure experimental techniques permit *in situ* measurements to be made of thermal conductivity under extreme conditions of pressure and temperature. The pressure-dependence of the conductivity of binary compounds with various mass-ratios was investigated[15] (table 4) by using a first-principles approach. This showed that compounds having a similar mass ratio (GaAs, SiC, BN, BP) exhibited an increased conductivity with pressure. Compounds having a high mass-ratio (BSb, BAs, BeTe, BeSe) and appreciable frequency-gaps between the acoustic and optical phonons exhibited a decrease in conductivity with pressure. The anomalous pressure-dependences were traced to the fundamentally different scattering processes obeyed by acoustic phonons. An abnormal strain-dependence has been observed in an hexagonal-phase monolayer BAs system. Calculations predicted the conductivity to increase from 180.2 to 375.0W/mK and 406.2W/mK along the armchair and zigzag directions, respectively, as a result of 3% of stretching. This enhancement was attributed stretching making the flexural out-of-plane mode the dominant heat-carrier.

Table 4. Thermal conductivities of various materials under pressure

Materials	Strain	Maximum Pressure	Thermal Conductivity (W/mK)
Bi_2Te_3	compressive	10GPa	2.47–5.6 (0–10GPa, 300K)
Sb_2Te_3	compressive	10GPa	1.22–3 (0–10GPa, 300K)
PbTe	compressive	11GPa	2–4 (0–11GPa, 300K)
PdS	compressive	10GPa	25–9 (0–10GPa, 300K)
$Pb_{0.99}Cr_{0.01}Se$	compressive	6GPa	2.1–8.2 (0–6GPa, 300K)
$CuInTe_2$	compressive	5GPa	7.5–4.1 (0–5GPa, 300K)
$CuInTe_2$	compressive	8GPa	2.1–8.2 (0–8GPa, 300K)
Si	torsion	24GPa	142–7.6 (0–24GPa, 300K)
Si	compressive	45GPa	73–300 (0–36GPa)
$Si_{0.991}Ge_{0.009}$	compressive	45GPa	24–360 (0–36GPa)
GaAs	compressive	20GPa	49–70 (0–16GPa, 300K)
BAs	compressive	80GPa	1331–823 (0–80GPa, 300K)
MoS_2	compressive	19GPa	3.5–25 (0–19GPa, 300K)

It was predicted[16] that many polytypes possessing the structural features of cubic and hexagonal diamond, which were known to be unstable even up to very high pressures, could become stable at low pressures and might be retained down to ambient conditions. Some of the polytypes were predicted to exhibit high thermal conductivities under ambient conditions. The thermal conductivities of zincblende and ^{2}H-BAs were expected to decrease with increasing pressure, with this being attributed mainly to stronger third anharmonic interactions.

Density functional perturbation theory was used[17] to simulate the 26 vibrational modes of $B_{12}As_2$ crystal at the Γ-point. All of the 12 Raman-active modes were in good agreement with the experimental Raman spectra. Dispersion curves were plotted from the Brillouin zone-center to the Z-point (0.5, 0.5, 0.5) and to the A-point (0.5, 0.0, 0.0). The acoustic-branch speeds were deduced to range from 9850 to 10980m/s for the LA modes and from 6940 to 7220m/s for the TA modes. Raman scattering spectroscopy was used[18] to study the spectral linewidth and frequency of the characteristic phonon modes of $B_{12}As_2$ as a function of temperatures of between 7 and 680K. A combination of 2- and 3-phonon decay processes dominated the dynamics of the intra-icosahedral phonon modes at 624, 682 and 742/cm, while 2-phonon decay was apparently favoured for the As-As stretching mode at 310/cm and the 506/cm mode. The changes in frequency with temperature involved mainly contributions which arose from lattice expansion, apart from the 506/cm mode where contributions which arose from phonon damping were more important.

By using molecular dynamics simulations which were based upon precise force-fields, it was shown[19] that phonon-scattering strengths which were produced by cubic anharmonicity could be greatly underestimated when applying the perturbation-theory approach to materials which possess large frequency-gaps or branch-bunching. This was attributed to the additional 3-phonon scattering in molecular dynamics models which was made possible by fluctuating phonon energy and continuous energy exchanges between modes. These were essential to the accurate evaluation of the zone-center phonon line-width in boron arsenide, as compared to experimental data and could appreciably decrease the lattice thermal conductivity of beryllium telluride and tungsten carbide. Due to the stronger 3-phonon scattering, 4-phonon scattering then became less important than was previously assumed.

The fixed mathematical form of the empirical potentials which are used in molecular dynamics simulations mean that many properties of materials cannot be predicted with experimental accuracy. Accurate quantum-theory electronic-structure calculations such as density functional theory are based meanwhile upon treating several hundred atoms within a picoseconds. This renders the method inadequate for treating systems beyond the nanoscale. The speed of classical molecular dynamics and the accuracy of density functional theory can be combined by using machine learning methods. Machine-learning offers powerful data-analysis capabilities[20] and has been successfully applied to many fields. Some particular applications in computational nanotechnology include property-prediction and material mining. Machine-learning potentials bridge the efficiency-versus-accuracy gap that exists between density functional calculations and classical molecular dynamics. In the case of property prediction, machine-learning provides a robust means that avoids any need for repetitive calculations for different simulation runs.

A so-called spatial density neural network force field method was developed[21] by training neural networks to predict forces at the density functional theory level. The model involved the use of a 3-dimensional mesh of density functions which worked together to map the atomic environment and provide a physical picture of the forces which acted upon the central atom. The advantages of the spatial density neural network force field include avoidance of the chain-rule by directly deducing the forces from the neural network. There was a marked reduction in the parameters and effort that was needed to train a force and/or property converged neural network force field. The spatial density neural network force field method was applied to the structures of bulk silicon, diamond, silicon carbide and boron arsenide. The phonon dispersions and lattice thermal conductivities could be predicted by using direct solutions to the phonon Boltzmann transport equation. In deducing the phonon properties a fitting method was used to obtain the second-order and third-order force constants. As compared with density functional theory calculations of the lattice thermal conductivity, higher-precision results were obtained by using the spatial density neural network force field; giving values which were within 0.7% in the case of silicon, 6.2% in the case of diamond, 2.76% in the case of SiC and 7.46% in the case of BAs. The phonon dispersions which were deduced by using the spatial density neural network force field method also matched those which were found by using direct density functional theory, or experiment.

It has also been noted[22] that a popular lowest-order theory of phonon-phonon interactions often fails to describe properly the anharmonic phonon decay rates and thermal conductivity, even for strongly bonded crystals. A first-principles theory which included the lowest-order 3-phonon and the higher-order 4-phonon processes was applied to 17 zincblende semiconductors. This revealed that the lowest-order theory markedly overestimated the measured conductivity of many of the materials, whereas inclusion of 4-phonon scattering provided a greatly improved agreement with measured data. New selection rules for 3-phonon processes helped to explain many of the above failures in terms of the anomalously low anharmonic phonon decay-rates which were predicted by the lowest-order theory when dealing with 4-phonon processes. Zincblende compounds which contained boron, carbon or nitrogen atoms exhibited a particularly weak 4-phonon scattering, as compared with that of compounds which did not contain those elements. This all helped to explain the ultra-high conductivities of materials such as cubic boron arsenide, boron phosphide and silicon carbide. It also cast doubt on the possibility of attaining high conductivities in materials which do not contain boron, carbon or nitrogen.

The photo-response of $B_{12}As_2$ bulk crystals to light of various wavelengths was studied[23], revealing that the crystals exhibited a photocurrent response to a band of 407nm and 470nm blue light. The maximum room-temperature photoresponsivity and photocurrent density at 0.7V for 470nm blue light were 0.25A/W and 2.47mA/cm^2, respectively. There was an exponential dependence upon temperature with regard to the dark-current, current density and resistivity. The latter parameters varied by a few nA to μA, by 1 to 100μA/cm^2 and by 7.6 x 10^5 to 7.7 x 10$^3\Omega$cm, respectively, at temperatures ranging from 50 to 320K.

A determination was made[24] of the coefficients of thermal expansion of semiconducting boron-rich icosahedral $B_{12}As_2$. Powder was synthesized in a sealed quartz ampoule which contained boron and arsenic that were heated to 1100C and 600C, respectively, for 72h. The lattice constants were measured by means of high-temperature X-ray diffraction at between 25 and 850C. The lattice parameters of the powder increased with temperature between 25 and 850C. The average lattice coefficients of thermal expansion were calculated, perpendicular to and parallel to the <111> axis in the rhombohedral phase, to be 4.9 x 10^{-6}/K and 5.3 x 10^{-6}/K, respectively. The average unit-cell volumetric coefficient of thermal expansion was 15.0 x 10^{-6}/K.

The volumetric and linear coefficients of thermal expansion were calculated from the lattice parameters. The coefficients of thermal expansion were plotted as a function of temperature. The relationship was fitted by using least-squares regression,

$$\alpha(/K) = \_\alpha_0 + \alpha_1 T + \alpha_2/T^2$$

(table 5). Here, α_0 was the most important term and described the linear dependence upon temperature, while α_2 described the curvature. The coefficient of thermal expansion was least sensitive to α_2 given that it was divided by the square of the temperature. There was some anisotropy between the axes; with a 10% difference in the coefficient of thermal expansion for the a and c axes.

When compared to bulk crystals or epitaxial films, $B_{12}As_2$ nanowires can have lower defect densities or be defect-free, leading to superior electrical properties and improved device performance. Nanowires were synthesized[25] by means of vapour-liquid-solid growth using platinum powder and nickel powder on silicon carbide and 20nm-thick nickel film on silicon substrates at 700 to 1200C. Platinum led to the best nanowires, from 900 to 950C. At these growth temperatures the diameters ranged from less than 30nm to about 300nm. Growth temperatures of 850C or less produced curled wires which were 200 to 1000nm in diameter. Excellent crystallinity was exhibited by wires which were grown at above 850C, whereas wires which were grown at, or below, 850C were partially amorphous. Wires which were grown from 20nm nickel film had similar morphologies for temperatures of up to 850C. From 900 to 950C, straight isolated wires were produced with diameters of 200 to 400nm. Nickel powder produced only wires which were more than 1μm in diameter.

Table 5. Coefficient of thermal expansion of $B_{12}As_2$ at 25 to 850C

Orientation	Coefficient	Value
a-axis	α_0	4.21×10^{-6}
a-axis	α_1	1.30×10^{-9}
a-axis	α_2	0.21
c-axis	α_0	4.9×10^{-6}
c-axis	α_1	1.2×10^{-9}
c-axis	α_2	0.2
volume	α_0	13.3×10^{-6}
volume	α_1	3.9×10^{-9}
volume	α_2	0.44

Crystals were grown[26] by precipitation from nickel metal solutions which were saturated with elemental boron, or $B_{12}As_2$ powder, and arsenic in a sealed quartz ampoule. Crystals which were 1015mm in size were produced when an homogeneous mixture of the three elements was held at 1150C for 4872h and cooled at a rate of 3.5C/h. The use of X-ray topography and energy-dispersive X-ray spectroscopy confirmed that the crystals had the expected rhombohedral structure and chemical stoichiometry. Etch-pit densities of only $4.4 \times 10^7/cm^2$ were found following etching in molten KOH at 500C, indicating the high quality of the crystals.

The crystallographic properties of bulk icosahedral boron arsenide crystals which were grown by precipitation from molten nickel solutions were characterized[27]. The large (5 to 8mm) crystals were produced by dissolving boron in nickel at 1150C for 48 to 72h, reacting with arsenic vapour and slowly cooling to room temperature. Raman spectroscopy, X-ray topography and defect-selective etching revealed that the single crystals were of high quality and contained low dislocation densities. The major face of the plate-like crystals was of (111)-type, while (100), (010) and (001)-type facets were also observed. The predominant defect was an edge-type growth dislocation with a <001> Burgers vector and a <$\bar{1}$10> line direction.

Defect-selective etching for the purpose of estimating dislocation densities in icosahedral boron arsenide crystals, by using molten potassium hydroxide, was applied[28] to crystals which had been produced from a molten nickel flux. The etching time ranged from 60 to 300s at between 400 and 700C. The etch-pits were mainly triangular and ranged in size from 5 to 25μm. The average etch-pit densities of triangle and oval etch-pits were of the order of 5 x $10^7/m^2$ and 3 x $10^6/cm^2$, respectively, for crystals that had been etched for 120s at 550C.

The epitaxial growth of icosahedral boron arsenide on 4H-SiC substrates that were misoriented from (00•1) and towards [1T•0] was shown[29] to eliminate rotational twinning. Synchrotron white-beam X-ray topography and high-resolution transmission electron microscopic comparisons of material which had been grown on off-axis and on-axis c-plane 4H-SiC confirmed the monocrystalline nature and far superior quality of films which were grown on the latter substrates. There was no intermediate layer between the epilayer and the substrate in the case of the off-axis 4H-SiC substrate. Vicinal steps that were formed by hydrogen-etching of the off-axis 4H-SiC surfaces before deposition caused the film to have a single orientation. This phenomenon was not observed when the substrate had no misorientation, or was tilted towards the [11•0] direction. It was concluded that c-plane 4H-SiC with a 7° off-cut towards [1T•0] was perhaps the optimum substrate for the growth of high-quality untwinned icosahedral boron arsenide epilayers. Thin films were deposited[30] onto 6H-SiC substrates by means of chemical vapour deposition using B_2H_6 and AsH_3 sources. X-ray diffraction analysis showed that the films had a rhombohedral crystal structure, with the lattice parameters of $B_{12}As_2$, and were polycrystalline with oriented crystal grains. The preferred orientation of the film with respect to the substrate was:

$$[00•1]_{B_{12}As_2} \| [00•1]_{6H\text{-}SiC}$$

and

$$[10•0]_{B_{12}As_2} \| [10•0]_{6H\text{-}SiC}$$

to within 3°. Electron diffraction examination also revealed the existence of an extremely small lattice mismatch of less than 0.5% between the $B_{12}As_2$ basal-plane lattice parameter and twice the SiC basal-plane lattice parameter.

The chemical vapour deposition of $B_{12}As_2$ onto on-axis and off-axis 6H-SiC (00•1) substrates was performed[31] by using hydrides as reactants. The growth rate increased with temperature, from 1.5μm/h at 1100C to 5μm/h at 1400C and then decreased. X-ray diffraction examination revealed that the deposits were amorphous when the deposition temperature was below 1150C. At higher temperatures, smooth $B_{12}As_2$ films formed on the 6H-SiC substrates, with

$$(00\bullet1)_{B12As2}||(00\bullet1)_{6H\text{-}SiC}$$

Raman spectroscopy confirmed a strong c-axis orientation of the arsenide film.

When thin epitaxial films of icosahedral boron arsenide were prepared[32] by chemical vapour deposition at 1300C onto on-axis 6H-SiC (00•1) substrates, using hydrides as reactants, synchrotron white-beam X-ray Laue patterns indicated that the epitaxial relationship was

$$(111)<101>||(00\bullet1)<11\bullet0>$$

with the presence of double-positioning twins of $B_{12}As_2$. Diffractometry further indicated that two twinning variants existed, in the ratio of 2:1. A reciprocal-space map of the epilayer revealed a single peak which was 12arcmin wide in the ω scan direction and 1.6arcmin wide in the θ-2θ scan direction. There was also a high degree of tilt, plus strain-broadening. Cross-sectional transmission electron micrographs revealed twin domains, with one twin orientation appearing as a matrix containing grains, between 0.2 and 4μm wide, of the second twin orientation. A high density of dislocations was also found.

The elimination of degenerate epitaxy during the growth of icosahedral boron arsenide on m-plane 15R-SiC substrates, and on 4H-SiC substrates which were misoriented by 7° from (00•1) and towards [1T•0], was studied[33]. Synchrotron white-beam X-ray topography showed that only a single orientation of the arsenide was present in the epitaxial layers. This confirmed the absence of the twin variants which were predominant signs of degenerate epitaxy. A low asterism of the arsenide diffraction-spots, as compared with those grown onto other SiC substrates, signalled a greater film quality. Cross-sectional high-resolution transmission electron microscopy and scanning transmission electron microscopy further confirmed the absence of twins in the films and their consequent high quality. The ease of nucleation on the ordered step structures which were present on these special substrates over-rode the symmetry considerations that encouraged degenerate epitaxy.

The thermal conductivity of icosahedral boron arsenide films which were grown onto (00•1) 6H-SiC substrates by chemical vapour deposition was studied[34] by using the 3ω technique. The room-temperature thermal conductivity decreased from 27.0 to 15.3W/mK when the growth temperature was decreased from 1450 to 1275C, and this was largely attributed to differences in the impurity concentrations and microstructures. Callaway's theory was used to calculate the temperature-dependent thermal conductivity, and the results were in good agreement with the experimental data. The Seebeck coefficients were 107 and 136μV/K for samples which were grown at 1350C by using AsH$_3$/B$_2$H$_6$ flow ratios of 1:1 and 3:5, respectively.

The resistivity of B$_{12}$As$_2$ crystals[35] had a nominal value of 15 to 18Ωcm at 400K and decreased to slightly above 4Ωcm at 700K. The electrical conductivity (figure 1) could be described by the growth of B$_{12}$As$_2$/SiC pn heterojunction diodes which were deposited onto (00•1) n-type 4H-SiC by chemical vapour deposition was studied[36]. The diodes exhibited a good rectifying behaviour, with an ideality-factor of 1.8 and leakage currents as low as 9.4 x 10^{-6}A/cm^2. Capacitance-voltage measurements which were performed by using a 2-frequency technique revealed a hole-concentration of 1.8 x 10^{17} to 2.0 x 10^{17}/cm^3 in B$_{12}$As$_2$. There was a slight increase near to the interface, due to the presence of an interfacial layer which accommodated the lattice mismatch. The band offsets between the B$_{12}$As$_2$ and SiC were estimated to be about 1.06 and 1.12eV for the conduction and valance bands, respectively.

Boron arsenide resembles silicon electronically to the extent that its Γ conduction-band minimum is p-like rather than s-like, and it has an indirect band-gap with its bond charge distributed almost equally on the two atoms in the unit cell; exhibiting almost perfect covalency. These aspects can be traced[37] to an anomalously low atomic p-orbital energy of boron and an unusually strong s-s repulsion in BAs; unlike most group-III-V compounds. Unexpected valence-band offsets of BAs are found, with respect to its GaAs and AlAs relatives. The valence-band maximum of BAs is much higher than that of AlAs regardless of the much shorter bond-length of BAs. The valence-band maximum of GaAs is only slightly higher than that of BAs. This results from the unusually strong mixing of cation and anion states at the valence-band maximum.

The high-pressure properties of the icosahedral phase were studied[38] by means of room-temperature *in situ* X-ray diffraction measurements performed at pressures of up to 25.5GPa. The phase retained its rhombohedral structure, and no phase transition was observed within the pressure range that was explored. The bulk modulus was 216GPa, with a pressure-derivative of 2.2. There was some anisotropy of the compressibility, in that the c-axis was 16.2% more compressible than was the a-axis.

$$\log[\sigma] = -490.87/T + 0.0941$$

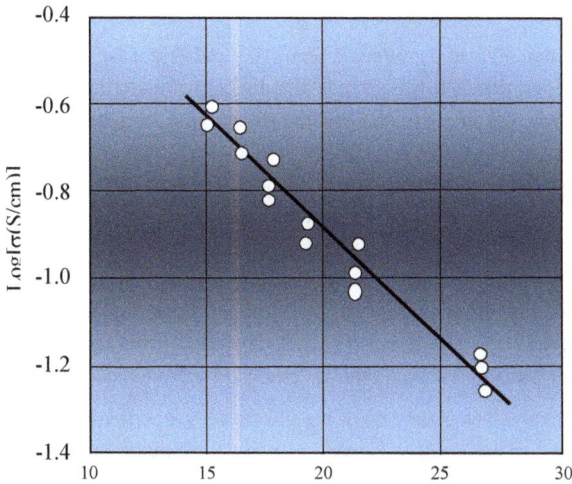

Figure 1. Conductivity of $B_{12}As_2$ crystals as a function of temperature

Hexagonal Boron Arsenide

Two-dimensional materials have become platforms for a wide array of nano-electronic devices. First-principles calculations and phonon interface transport modelling were used[39] to calculate the temperature-dependent thermal boundary conductance of single layers of graphene-analogue two-dimensional materials such as silicene, hexagonal-BN, boron arsenide and blue or black phosphorene on amorphous SiO_2 or crystalline GaN substrates. These results showed that, in 2-dimensional/3-dimensional systems, the room-temperature thermal boundary conductance could range from 7 to 70MW/m²K. The lowest value here belonged to BP, and the highest to hexagonal-BN. The 2-dimensional/3-dimensional thermal boundary conductance exhibited a marked temperature dependence, but this could be reduced by encapsulating the 2-dimensional/3-dimensional assembly. Encapsulation with AlO_x caused the thermal boundary conductance of some graphene-analogue 2-dimensional materials to equal or exceed those (15 to 40MW/m²K) of graphene and transition-metal dichalcogenides. The room-temperature thermal boundary conductance was calculated as a function of the van der Waals spring coupling, where the thermal boundary conductance then ranged from 50 to 150MW/m²K for coupling strengths of 2 to 4N/m in the case of silicene, BAs and blue phosphorene. Group III-V materials having ultra-soft flexural branches were proposed to be promising 2-dimensional materials for thermal isolation when matched to crystalline substrates.

Hexagonal boron arsenide is a graphene analogue and is among the most attractive 2-dimensional semiconductors. It was found[40] that this novel material can possess an elastic modulus of 260GPa, regardless of the loading direction. The material also has a narrow direct and band-gap of 1.0eV. The thermal conductivity of a stress-free pristine monolayer was predicted to be 180.2W/mK, and this could be increased to 375.0 and 406.2W/mK, by just 3% of straining along the armchair and zig-zag directions, respectively. The mechanism for this increase in thermal conductivity was related to the fact that stretching made the flexural out-of-plane mode the dominant heat-carrier. Hybrid density functional calculations for single-layer boron arsenide showed[41] that the material had a direct band-gap of 1.2eV which originated from the energy difference of the p z orbitals and was tunable by straining. The band-gap increased non-linearly.

A first-principles analysis was made[42] of the electronic properties of single-walled boron arsenide zig-zag nanotubes. Local density approximation, Perdew-Zunger parameterization and generalized gradient approximation under Perdew Burke Ernzerhoff were used as exchange correlation functionals. The generalized gradient approximation results confirmed that the BAs nanotubes were metallic up to some critical diameter and then exhibited semiconducting behaviour. The structural, mechanical and transport properties of boron arsenide nanowires of various sizes, having an hexagonal cross-section, were again analyzed[43] by using density functional theory based upon a first-principles approach. The generalized gradient approximation, with a revised Perdew-Burke-Ernzerh-type parameterization, was applied. The stability was analyzed in terms of binding energies by optimizing the structures with regard to lattice parameters, using a total energy minimization approach. Arsenide nanowires with larger cross-sections were found to be more stable among the size-range considered. The variation in the conducting ability of the nanowire as a function of the diameter was studied by means of band-structure and density-of-states calculations. The mechanical and transport properties were also studied for various cross-sections of the hexagonal BAs nanowire. The electronic structure and carrier mobility of single-walled boron arsenide nanotubes were studied[44] using the Boltzmann transport equation with the relaxation-time approximation. The ionic character of the B-As bond resulted in the presence of dipole shells in the optimized nanotubes. It was predicted that both zig-zag and armchair nanotubes would be semiconductors. Strong σ^*-π^* hybridization in small zig-zag nanotubes leads to a rapid decrease in band-gap with decreasing radius. As the size of the nanotube decreases the hole-mobility in zig-zag nanotubes exhibits a 3p (p an integer) oscillation but the electron mobility decreased, and did so even more rapidly when the radius decreased. It was concluded that the rolling of nanoribbons into nanotubes would increase the electron-mobility but decrease the electron-mobility. The differing behaviours of the carrier mobility in nanoribbons and nanotubes were due to the distinct bond features of edge states, which were a function of the width for nanoribbons or the radius for nanotubes.

By solving the Boltzmann transport equation it was shown[45] that 4-phonon scattering has a significant effect on thermal transport in honeycomb-structured monolayer BAs, m-BAs, and in its hydrogenated bilayer counterpart, bi-BAs. The lattice thermal conductivities of all these structures are however reduced upon incorporating 4-phonon scattering: a drop in the conductivity of up to 80% is found for m-BAs when compared to the absence of 4-phonon scattering, which is mainly caused by the suppression of phonon lifetimes. The conductivity of m-BAs is anomalously lower than that of its bi-BAs counterpart and this attributed to the much higher phonon scattering-rate in m-BAs, as compared to that in bi-BAs. Upon comparing BAs sheets, with and without horizontal mirror symmetry, it is furthermore found that the contribution from flexural acoustic phonons suffers its greatest reduction in both m-BAs and bi-BAs with horizontal mirror symmetry upon including 4-phonon scattering.

The conductivity of honeycomb-structured layered monolayer boron arsenide can attain 181W/mK at 300K, due to the unusual phonon lifetime. The conductivity of hydrogenated bilayer BAs is much larger due to the formation of covalent bonds. The effect of horizontal mirror symmetry plus high-order phonon-scattering upon thermal transport in honeycomb-structured layered BAs remains unclear. A was thus made of m-BAs and of 2 hydrogenated bi-BAs samples with the incorporation of 4-phonon scattering. At 300K, the conductivity of m-BAs was reduced from 176 to 35W/mK when 4-phonon scattering was included. In the case of the bi-BAs samples, 4-phonon scattering caused a reduction of less than 30%, while the hydrogenated bilayer structures maintained conductivities of 223 and 166W/mK at 300K. The conductivity of m-BAs was lower than that of bi-BAs because of the larger phonon-scattering rate. By comparing bi-BAs, with and without horizontal reflection symmetry, the effect of horizontal reflection symmetry upon conduction could be examined. The m-BAs and bi-BAs-AA possessed horizontal mirror symmetry, but this was absent from bi-BAs-AB. The lattice constants, following relaxation were 3.39, 3.42 and 3.43Å for m-BAs, bi-BAs-AA and bi-BAs-AB, respectively. The effective thickness is conventionally defined as the sum of the buckling-height and the van der Waals radii of the outermost atoms. The effective thickness values are therefore 4.26, 8.11 and 8.20Å for m-BAs, bi-BAs-AA and bi-BAs-AB, respectively. These are then used in the calculating the conductivity. Due to the mirror symmetry induced selection rule in m-BAs and bi-BAs-AA, some phonon-scattering channels which involve ZA phonons are forbidden in m-BAs and bi-BAs-AA, but this restriction disappears in bi-BAs-AB due to the absence of mirror symmetry. The calculated 3-phonon limited conductivity of m-BAs was 176W/mK at 300K. It is known that the conductivity of cubic BAs may be decreased by some 48% at 300K upon including 4-phonon scattering. The present results confirmed that 2-dimensional BAs also exhibits a decreased conductivity upon including 4-phonon scattering. The reduction in conductivity by 4-phonon scattering was much smaller (29% and 28%, respectively) for bi-BAs-AA and bi-BAs-AB. Its effect was nevertheless not negligible. Covalent bonds are formed in bi-BAs, which then improve the mechanical properties of the structure. The conductivity of bi-BAs-AA was 6 times greater than that of m-BAs upon including 4-phonon scattering. Due to the mass difference between boron and arsenic atoms, the phonon dispersion in m-BAs could be separated into low-frequency and high-frequency regions, with a large band-gap (13THz) between them. Upon adding H-passivation to bilayer BAs sheets the flat band at ultra-high frequencies appeared due to the low mass of the hydrogen atom. Additional phonon branches existed in the phonon dispersion of bi-BAs-AA and bi-BAs-AB. These were induced by the interlayer covalent bonds.

Two-dimensional stable monolayers of boron arsenide sheet were investigated[46] by using state-of-the-art theoretical calculations. The band-gap of the semiconducting sheet changed from direct to indirect upon applying a biaxial strain of 10%, and it became metallic at 14% of biaxial strain. Positive phonon vibrational modes were observed for all of the applied biaxial strains, ensuring the stability of the sheet under strain. The semiconducting property was retained upon cutting 2-dimensional sheet into 1-dimensional nanoribbons, and the band-gap was size-dependent. The calculated optical properties exhibited a strong anisotropy. The BAs nanomaterial exhibited a marked adsorption in the ultraviolet to visible region. The calculated Seebeck coefficient and power-factor suggested that BAs sheet was an ideal candidate for thermal management and thermo-electric applications. The thermodynamic properties were calculated on the basis of the phonon frequencies.

A survey has presented[47] extensive tables of the key thermoelectric properties which define thermoelectric conversion efficiency for a wide range of materials. The materials which were included in these tables were selected mainly on the basis of well-established and internationally-recognized performance. Because thermoelectric properties vary with temperature, the data are given for ambient temperatures in order to permit easy comparison, and for higher temperatures linked to peak performance. The types of material covered are tellurides, skutterudites, half-Heuslers, Zintls, antimonides, clathrates, $FeGa_3$-type, actinides, lanthanides, oxides, sulfides, selenides, silicides, borides and carbides. Boron arsenide has not yet found a place on this list

A systematic density functionial theory study was made[48] of 2-dimensional BAs with regard to the effect of inherent defects upon the electronic structure, and magnetic and thermoelectric properties. The defect in which an As atom replaced a B atom maintained the semiconducting properties while all other defects caused the material to exhibit metallic properties. Vacancy and interstitial defects introduced magnetism; originating mainly from the s and p orbitals of the atoms near to the defects. A novel observation was that a B atom which replaced an As atom markedly improved the thermoelectric performance. The typical defects which are found in 2-dimensional materials are vacancy defects, antisite defects and interstitial defects. Common defects in 2-dimensional BAs are vacancy defects involving a single arsenic atom, V_{As}, vacancy defects involving a single B atom, V_B, double vacancy defects involving a B atom and an As atom, V_{BAs}, a defect with an extra As atom in the gap, As^+, a defect with an extra B atom in the gap, B^+, an antisite defect involving a B atom replacing an As atom, B_{As}, and an antisite defect involving an arsenic atom replacing a B atom, As_B. By using a rigid model combined with semi-classical Boltzmann theory, the effect of these seven defects upon the thermoelectric properties was determined at 300 to 2500K. These properties included electrical conductance, Seebeck coefficient, phononic thermal conductivity and ZT; a measure of the thermal conductivity. Thermoelectrics generate power when located in a temperature gradient, or cool when a current is passed through the material. The thermoelectric performance in either mode depends upon the efficiency of the material in converting heat into electricity. The efficiency of a thermoelectric material depends mainly upon the thermoelectric material's figure-of-merit: known as ZT. In its simplest form ZT is described by:

$$ZT = S^2\sigma T/\kappa$$

where the voltage generated is defined by the Seebeck coefficient, S, In order to maximize the efficiency at a particular temperature a high electrical conductivity, σ, is required, together with a low thermal conductivity, κ. The latter is in turn made up of 2 components, the lattice thermal conductivity and the electronic thermal conductivity. The electrical transport and the electronic contribution to thermal transport are directly linked via the Wiedemann–Franz law. There is thus great interest in modifying σ and κ independently in order to maximize ZT.

Most defect types significantly increased the electrical conductivity. The B_{As} defect could increase the conductivity by a small amount when the temperature was greater than

800K, but the V_B defect caused the conductivity to decrease slightly. With increasing temperature, the conductance of B_{As}, B_{As} and B^+ exhibited an increasing trend. The V_{As} defect led to a small increase and then a decrease. The other defects caused a marked overall decrease. Various defects increased the phononic thermal conductivity to some extent, and the temperature had a very slight effect upon its value. The effects of As^+ and V_{As} upon the phononic thermal conductivity were very similar. When the temperature reached some 1000K, the peak value of ZT was 0.78. This was attributed to the replacement of arsenic atoms by boron atoms, which reduced the mass ratio of the system. The small ratio decreased the frequency between the acoustic and optical phonons. Meanwhile, the 3-phonon scattering and the higher-order phonon scattering were weakened.

Density functional theory calculations were used[49] to study pristine and hybrid bilayers of hexagonal boron phosphide and hexagonal boron arsenide. The electronic and optical properties of the bilayers could be modulated by applying an external perpendicular electric field. The band-gaps could then be tuned from 0.8eV to zero as the field was increased up to a critical value. At above this value the gap re-opened because the valence and conduction bands underwent an anti-crossing of so-called Mexican-hat form; similar to that observed in biased graphene bilayers. There was an intense peak in the absorption spectra at 2.5 to 2.6eV, with a slight blue-shift as a function of the electric field. At above the critical field value, a new peak was found in the infra-red region, and this exhibited a strong field-dependence. This peak was related to optical transitions occurring around the Mexican-hat region of the band structure. The corresponding electron-hole pair exhibited a layer separation which could lead to larger recombination times.

The structural and electronic properties of Janus-like boron arsenide bilayers were investigated[50] with the aid of density functional theory. One hydrogenated and one fluorinated BAs layer were combined in various stacking arrangements, and the interlayer charge transfer was predicted under the assumption of divers charge-population mechanisms. The choice of the basis set was found to be a critical factor in choosing the H-F interlayer distances (table 6) and therefore the energy-gaps (table 7). A Bader charge analysis (table 8) revealed charge accumulations around the fluorine and hydrogen atoms which were consistent with their electronegativity values, with an accumulation of charge-population around fluorinated layers. The layer-resolved projected density of states at the van der Waals density functional level revealed that the bilayers were type-II semiconductors, given that the valence and conduction bands were composed of different layers. The band-gap (table 9) of the bilayers could be varied from 0 to 0.54eV by controlling the layer arrangement. The electronic properties could also be varied from semiconducting to semimetallic.

Optimized crystal structures, energy-band structures and density of states data were investigated[51] by using full-potential linearized augmented plane-wave methods. The exchange-correlation potential was treated by using the generalized gradient approximation. The predicted results, such as band-gaps, were in reasonable agreement with a reported figure of 1.2eV.

Table 6. Interlayer H::F distance determined for bilayers using various models

Bilayer	Model	Interlayer Distance (Å)
HAsBH-FAsBF	PBE-D2	2.51
HAsBH-FBAsF	PBE-D2	2.53
HBAsH-FAsBF	PBE-D2	2.48
HBAsH-FBAsF	PBE-D2	2.52
HAsBH-FAsBF	PBE-D2	2.56
HAsBH-FBAsF	PBE-D2	2.56
HBAsH-FAsBF	PBE-D2	2.5
HBAsH-FBAsF	PBE-D2	2.51
HAsBH-FAsBF	vdW-DF	2.49
HAsBH-FBAsF	vdW-DF	2.6
HBAsH-FAsBF	vdW-DF	2.42
HBAsH-FBAsF	vdW-DF	2.55
HAsBH-FAsBF	vdW-DF	2.53
HAsBH-FBAsF	vdW-DF	2.53
HBAsH-FAsBF	vdW-DF	2.4
HBAsH-FBAsF	vdW-DF	2.48
HAsBH-FAsBF	M06-L6-311G*	2.84
HAsBH-FBAsF	M06-L6-311G*	2.84
HBAsH-FAsBF	M06-L6-311G*	2.83

HBAsH-FBAsF	M06-L6-311G*	2.87
HAsBH-FBAsF	M06-L6-311G*	2.80
HAsBH-FBAsF	M06-L6-311G*	2.91
HBAsH-FAsBF	M06-L6-311G*	2.86
HBAsH-FBAsF	M06-L6-311G*	2.91

Table 7. Relative energies for bilayers using various models

Bilayer	Model	Relative Energy (meV/atom)
HAsBH-FAsBF	PBE-D2	2.950
HAsBH-FBAsF	PBE-D2	3.116
HBAsH-FAsBF	PBE-D2	0.00
HBAsH-FBAsF	PBE-D2	1.940
HAsBH-FAsBF	PBE-D2	3.186
HAsBH-FBAsF	PBE-D2	3.006
FBAsF-HAsBH	PBE-D2	1.347
FAsBF-HAsBH	PBE-D2	2.827
HAsBH-FAsBF	M06-L	2.140
HAsBH-FBAsF	M06-L	2.14
HBAsH-FAsBF	M06-L	0.00
HBAsH-FBAsF	M06-L	1.94
HAsBH-FAsBF	M06-L	2.17
HAsBH-FBAsF	M06-L	1.36
FBAsF-HAsBH	M06-L	2.38
FAsBF-HAsBH	M06-L	1.73

Table 8. Band-gaps of HAsBH-FAsBF bilayers using various models

Bilayer	Model	Band-Gap (eV)
HAsBH-FAsBF	PBE-D2	-0.33
HAsBH-FBAsF	PBE-D2	-0.14
HBAsH-FAsBF	PBE-D2	-0.47
HBAsH-FBAsF	PBE-D2	-0.44
HAsBH-FAsBF	PBE-D2	-0.31
HAsBH-FBAsF	PBE-D2	-0.08
HBAsH-FAsBF	PBE-D2	-0.45
HBAsH-FBAsF	PBE-D2	-0.44
HAsBH-FAsBF	vdW-DF	-0.07
HAsBH-FBAsF	vdW-DF	0.18
HBAsH-FAsBF	vdW-DF	-0.90
HBAsH-FBAsF	vdW-DF	-0.22
HAsBH-FAsBF	vdW-DF	-0.03
HAsBH-FBAsF	vdW-DF	0.33
HBAsH-FAsBF	vdW-DF	-0.72
HBAsH-FBAsF	vdW-DF	-0.20

Table 9. Bader charge numerical analysis calculated at the DFT-D2 theory level

Bilayer	H-B	B	As	H-As	F-B	B	As	F-As
HAsBH-FAsBF	2.0921	1.9029	3.9689	2.0036	8.01	2.0417	3.9999	7.981
HAsBH-FBAsF	2.0961	1.9054	3.9648	2.0038	8.0258	2.0097	4.0289	7.9654
HBAsH-FAsBF	2.1109	1.9053	3.9176	1.9999	8.0142	2.0542	4.0164	7.9815
HBAsH-FBAsF	2.1177	1.875	3.9706	1.9932	8.0269	2.0322	4.0204	7.9641
HAsBH-FAsBF	2.1073	1.8856	3.9567	2.017	8.0136	2.0115	4.0267	7.9817
HAsBH-FBAsF	2.1177	1.9208	3.9235	1.9952	8.0247	2.0399	4.0112	7.9669
HBAsH-FAsBF	2.1151	1.9089	3.9071	2.0016	8.0133	1.984	4.0837	7.9862
HBAsH-FBAsF	2.1098	1.8852	3.9664	2.0088	8.0276	2.0093	4.0322	7.9607

First-principles density functional theory calculations were also used[52] to explore the possibility of using hexagonal boron-arsenide as an anode material for alkali-based batteries. It was found that the adsorption strength of alkali atoms (lithium, sodium, potassium) on the arsenide, when compared with graphene and other materials, changed only a little as a function of the alkali-atom concentration. When the separation between the alkali atoms and the arsenide was less than a critical distance of about 5Å, the adsorption energy suddenly increased; thus indicating the occurrence of rapid adsorption without an energy barrier. The low energy-barriers of 0.322, 0.187 and 0.0.095eV for lithium, sodium and potassium, respectively, guaranteed fast ionic diffusivies for all three alkalies. The addition of these alkali atoms also transformed the electronic properties of the arsenide from semiconducting to metallic, thus resulting in improved electronic conductivities. The excellent storage capacity, of about 626mAh/g, of the arsenide for alkali atoms made it a competitor to other anode materials. State-of-the-art density functional theory calculations were similarly used[53] to explore the use of 2-dimensional honeycomb boron arsenide as an anode for alkali metal ion batteries. Structural and dynamic stability was confirmed by the formation energy and the non-negative phonon frequency. The arsenide monolayer had negative adsorption-energy values of -0.422, -0.321 and -0.814eV for lithium, sodium and potassium ions, respectively. During charging, the adsorption-energy increased appreciably with no energy-barrier when any of the alkali atoms reached a critical distance of about 8Å. Insertion of an alkali metal atom into the arsenide surface caused the semi-conducting nature of the monolayer to transform into a metallic-state. The low-energy barriers for lithium (0.522eV), sodium (0.248eV) and potassium (0.204eV) ion migration implied the occurrence of rapid diffusion over the arsenide surface. This suggested that it possessed a high charge/discharge capability. Low average operating voltages of 0.49V (lithium), 0.35V (sodium) and 0.26V (potassium) were found, together with high theoretical capacities of 522.08mAh/g (lithium, sodium) and 209.46mAh/g (potassium).

Group theory and density functional theory methods were combined[54] in order to obtain compact and accurate k•p Hamiltonians which could describe the band structures around the K- and T-points of 2-dimensional hexagonal boron arsenide. The latter is a direct band-gap material with band extrema at the K-point. The band-gap becomes indirect with a conduction-band minimum at the T-point when subjected to a strong electric field or biaxial strain. At high (circa 0.75/V) electric field strengths or a large (14%) strain, the 2-dimensional hexagonal boron arsenide becomes metallic.

The fundamental band-gap at the K-point was 0.76eV, but this transition was not allowed in the dipole approximation. The conduction band at the Γ-point was very sensitive to strain and electric fields, which made a transition to a metallic state possible[55].

An electrically tunable quantum spin Hall insulator with a large gap was created[56] in the van der Waals heterobilayer of a monolayer transition metal dichalcogenide and hexagonal boron arsenide; in particular, the WSe$_2$/BAs heterobilayer[57]. When the type-II band alignment was inverted in an electric field, the hybridization by interlayer hopping between the spin-valley locked valence band edges in the transition metal dichalcogenide and the BAs conduction band edges led to a stacking-configuration dependent topological band inversion. In the non-interacting limit, the double spin degeneracy of BAs left an un-hybridized conduction band within the gap, so the heterobilayer was a spin-valley locked metal instead of a quantum spin Hall insulator. Given the Coulomb interaction in the double-layer geometry, interaction with the hybridization-induced electric dipole shifted this un-hybridized conduction band upwards in energy, thus giving rise to an appreciable global quantum spin Hall gap. In a long-period moiré pattern with spatial variation of local stacking configurations, competition between Coulomb interaction and interlayer hopping led to superstructures of quantum spin Hall insulators and excitonic insulators.

Various gaseous atmospheres (N$_2$, O$_2$, CO$_2$, H$_2$O, CO, NO, NO$_2$, NH$_3$, SO$_2$), absorbed on pristine hexagonal boron arsenide were analysed[58] using density functional theory methods, with various adsorption positions being considered for each molecule. The most stable adsorption depended upon position, adsorption energy, charge transfer and work function. The SO$_2$ molecules had the best adsorption energy and a certain amount of charge transfer.

The effect of 0-dimensional vacancy defects (single boron vacancy, double boron vacancy, single arsenic vacancy, double arsenic vacancy, single boron-single arsenic vacancy) upon the structural, electronic, magnetic and optical properties of hexagonal boron arsenide monolayers was investigated[59] by using first-principles calculations. Density functional theory calculations indicated the existence of arsenic-vacancy induced magnetism (1.0µB) in the boron atoms around the vacancy site. Induced magnetism was absent in the case of the other vacancy defects. The magnitude of the induced magnetic moment for the arsenic vacancy decreased with the increasing vacancy concentration. A semiconductor-to-metal transition occurred due to the introduction of a single boron vacancy, a double boron vacancy or a double arsenic vacancy. A single boron–arsenic vacancy could decrease the pristine band-gap although a finite band-gap remained. Introduction of a 0-dimensional vacancy modified the optical absorption spectra of the monolayer. Work-function calculations showed that it increased when a vacancy defect was introduced.

It is interesting to compare the properties of this 2-dimensional arsenide with other 2-dimensional materials. The rule-of-thumb that lone-pair electrons lead to a low thermal conductivity has been used as a guide to what has to be done in order to promote that conductivity. It has been shown[60] that the conductivity can in fact be increased by up to 2 orders-of-magnitude by activating lone-pair electrons. Boron atoms were introduced into 2-dimensional graphene analogues such as phosphorene, arsenene, antimonene and boron arsenide (table 10) in order to activate lone-pair electrons. The anomalous results arose partly from an abnormal increase in relaxation time. The thermal conductivities of the materials (table 11) were calculated for 300K by using various methods. The iterative thermal conductivity of the 2-dimensional phosphorene, arsenene and antimonene was 16.86, 3.76 and 0.69W/mK, respectively. These values could be up to 4 orders-of-magnitude lower than the thermal conductivity of graphene (3094W/mK). The boron atom replaced an atom of the primitive cell and established a lone-pair electron model with supposedly lower thermal conductivity. Anomalously high thermal conductivity of 228.03, 138.59 and 21.85W/mK, respectively, were found. The phenomenon was explained in terms of competitive effects of phonon thermal transport between in-plane and out-of-plane acoustic phonon modes, between Grüneisen parameters and scattering phase space and between unbonded s-electrons. The failure of lone-pair electrons to lead here to a low thermal conductivity was attributed to the formation of a planar structure when the lone-pair electrons were activated. The delocalized electrons were then relatively evenly distributed within the space of the planar structure. The phenomenon was ultimately related to the structure of low-dimensional materials because that structure directly modulated the electron distribution.

Table 10. Group velocities of ZA, TA and LA phonon modes
in the center of the Brillouin zone of various graphene-like materials

Property	BAs	Antimonene	Phosphorene	Arsenene
ZA (km/s)	0.57	0.39	0.68	0.83
TA (km/s)	5.57	2.12	5.56	3.11
LA (km/s)	9.43	3.26	8.11	4.57
C_{11}	127.394	32.45	79.15	52.75
C_{12}	35.66	6.55	8.56	9.51
C_{66}	45.87	12.95	35.3	21.62
E (MPa)	117.41	31.13	78.22	51.03
μ	0.28	0.2	0.11	0.18

Table 11. Lattice constant, bond angle, effective thickness, maximum frequency, Debye temperature, Grüneisen parameter and lattice thermal conductivity of BAs and graphene analogues

Property	BAs	Phosphorene	Arsenene	Antimonene
A (Å)	3.39	3.28	3.61	4.12
α (°)	120	92.91	91.97	90.83
h(Å)	3.7	4.3	4.46	5
ω_{max}(THz)	25.25	15.94	8.97	6.02
θ (K)	1211.93	765.06	430.38	288.79
γ	-0.38	-2.17	-10.19	-2.87
κ_L (W/mK)	138.59	16.86	3.76	0.69

It is a matter of some importance whether the high conductivity persists when the structure changes from three-dimensional to graphene-like two-dimensional. Previous studies had investigated conductivity only with regard to 3-phonon scattering and isotope scattering, and yielded various results. Calculations were first made here[61] of second-order interatomic force constants and third-order interatomic force constants in order to solve iteratively the Boltzmann transport equation and thus obtain the conductivity of monolayer hexagonal compounds while considering only 3-phonon and isotope scattering. The corresponding conductivity was 205.7W/mK at room temperature. A monotonic decrease in conductivity occurred with increasing average atomic mass if arsenic was replaced by phosphorus or antimony. By calculating the fourth-order interatomic force constants it was possible to obtain the conductivity of monolayers, with the inclusion of 4-phonon scattering. The conductivity at room temperature was 37.99W/mK, which was very consistent with the conductivity of monolayers of the hexagonal material; as predicted by using the phonon spectral energy density method. This method considered all-order scattering and gave a value of 18.20mK. The results showed that the effect of 4-phonon scattering upon the conductivity of monolayer hexagonal BAs was considerable, and that the conductivity changed upon including 4-phonon scattering. The lattice conductivity of semiconductors can be estimated from the phonon dispersion curve, because the dispersion-curve slope reflects the phonon-group velocity. The dispersion of phonon branches implies that the number of 3-phonon scattering channels which dominate the conductivity is small. The dispersion curve of monolayer hexagonal BAs was relatively flat, due to its higher average atomic weight. The super-high conductivity of the cubic material was attributed to the large band-gap between the acoustic and optical branches, and to aggregation of the acoustic branches. Those factors no longer existed in monolayers of the hexagonal BAs. With regard to the dispersion curve, there was a large band-gap between the LO and TO branches and other acoustic modes, but the ZO branch was unique and existed close to the LA and TA branches. This fact led to the supply of more scattering channels for acoustic and optical phonons; thus resulting in a low conductivity. The ZO branch in monolayer hexagonal material is located in the mid-frequency region, just as in a honeycomb structure. The low-frequency acoustic branch had no associated aggregation phenomenon when compared with the cubic material. The acoustic phonons then enjoyed more opportunities for participation in the phonon-scattering process.

Table 12. Average atomic mass (mavg) of monolayer hexagonal materials

Material	mavg (amu)	Energy Cut-Off (eV)	a (Å)	Bond-Length (Å)
BP	20.89	650	3.213	1.855
BAs	42.87	550	3.392	1.958
BSb	66.29	600	3.739	2.159

There exists a relationship between the monotonic change in the group velocity, the bond-length and the average atomic mass (table 12). The bond-length, d, reflects the strength, V, of orbit coupling, such that

$$V \sim 1/d^2$$

An increase in bond length reduces orbit overlapping and this decreases the bond strength, thus determining the phonon velocity. A greater value of mavg implies a low vibration frequency, thus reducing the group velocity. The dependence arose from the fact that atoms with a mass, M, are connected by bonds with a spring-constant, K, leading to,

$$\text{phonon-velocity} \sim (K/M)^{0.5}$$

With regard to the 3-phonon scattering phase space, the monotonic variation is due to the fact that the phonon branches below the band-gap gradually get closer with increasing mavg and atomic mass ratio, thus providing more paths for the 3-phonon scattering process. If the energy difference between phonons is large, energy conservation will not be easy to satisfy.

The intensity of the scattering process also influences the scattering rate and the Grüneisen parameter reflects the phonon-scattering intensity; i.e. anharmonicity. The anharmonicity of each material can be estimated by finding the total Grüneisen parameter value. The absolute value of the total Grüneisen parameter of hexagonal BAs was 3.94, as anticipated from the monotonic change with increasing mavg increase. Strong anharmonicity reduces the conductivity. The number of 3-phonon scattering channels, and the scattering intensity of the monolayer phase collectively determine the monotonic change in phonon lifetime. Within most frequency-ranges, the phonon lifetime decreases with increasing mavg. The phonon lifetime of hexagonal BAs is markedly higher only within the 3.4 to 7THz band.

Measurements showed that the conductivity decreases with increasing temperature, via an inverse relationship. This was due mainly to an increase in phonon-phonon scattering that was caused by the increase in temperature. Because the relaxation-time approximation treats scattering as a pure thermal resistance process, it underestimates the conductivity. The room-temperature conductivity of hexagonal BAs with a thickness of 4.26Å was 205.7W/mK. The conductivity of III–V boron compounds, with regard to 3-phonon scattering, tends to exhibit a non-monotonic behavior as mavg increases. The present results revealed a monotonic change in the conductivity, phonon group-velocity and phonon lifetime as a function of mavg.

The thermal conductivity of monolayer honeycomb BAs is much lower than the bulk material. On the basis of Boltzmann's transport equation and non-equilibrium Green's function calculations it has been suggested[62] that covalently bonded bi-layer BAs can exhibit a thermal conductivity which is about twice that of bulk silicon. The thermal conductivity of bi-layer BAs was about 70% higher than that of the monolayer form. This novel behaviour depended upon largely suppressed phonon-scattering phase-space and anharmonicity. Because there are 7 atoms in each unit cell, there are 3 acoustic branches and 18 optical phonon branches. The lack of an imaginary frequency mode in the phonon dispersion indicates that these bi-BAs structures are dynamically stable (table 13). The out-of-plane flexural acoustic branch of As-As bonding structures is flat and has no entanglement with other phonon branches. This is similar to that in hexagonal BAs. In the mid-frequency region of 10 to 30THz the phonon branches in As-As bonding structures are bunched together and lead to a wide band-gap in the phonon spectrum. The phonon group-velocities of in-plane transverse and longitudinal acoustic modes near to the G-point were listed (table 14). In bi-BAs both the v_{TA} (4.97 to 5.12km/s) and v_{LA} (7.64 to 7.84km/s) were slightly below those (v_{TA} = 5.54km/s, v_{LA} = 9.42km/s) of hexagonal BAs but close to those (v_{TA} = 4.95km/s, v_{LA} = 7.54km/s) of cubic BAs. This was attributed to the similar sp^3-hybridized bonds in bi-BAs and cubic BAs. The mechanical properties satisfied the Born-Huang mechanical stability criteria for 2-dimensional hexagonal crystals: that C_{11} > C_{12} and C_{66} > 0. This confirmed that bi-BAs allotropes are mechanically stable. The calculated 2-dimensional Young's moduli (164 to 175N/m) of bi-BAs allotropes were about 40% larger than that of hexagonal BAs. The Poisson ratios of 0.156 to 0.177 were about 40% smaller; suggesting that bi-BAs allotropes have superior mechanical properties. Electronic band structure calculations indicated that bi-BAs structures are semiconductors in which phonons govern thermal transport, because of the low carrier concentration. By using the 2nd and 3rd interatomic force constants calculations were made of the phonon thermal conductivity of bi-BAs. The temperature-dependent thermal conductivity of the two most energetically stable bi-BAs allotropes and of hexagonal BAs were calculated (figure 2). The thermal conductivity of the As-As structure could attain 313W/mK at room temperature; 72.9% higher than that of hexagonal BAs. The conductivity of bi-BAs was nevertheless lower than that of zincblende BAs and other bulk BAs samples. The lattice thermal conductivity is affected by the phonon group velocity and the phonon scattering rate. The phonon group velocities of bi-BAs allotropes were slightly lower than that of h-BAs. The main factor which enhanced thermal conductivity was therefore the differing anharmonic natures of h-BAs and bi-BAs allotropes.

Table 13. Space group, lattice constant, effective thickness, cohesive energy and formation energy of bi-BAs allotropes

	AA(B-As)	AB(B-As)	AA(B-B)	AB(B-B)	AA(As-As)	AB(As-As)
Space group	P3m1	P3m1	P6m2	P3m1	P6m2	P3m1
a(Å)	3.41	3.44	3.39	3.41	3.42	3.43
Thickness (Å)	8.14	8.47	8.25	8.17	8.11	8.20
E_{coh} (eV/atom)	3.820	3.823	3.751	3.776	3.884	3.880
E_f (eV/atom)	0.110	0.113	0.041	0.066	0.174	0.170

Table 14. Phonon group velocity of TA and LA branches at the G-point, elastic constants, in-plane Young's modulus and Poisson ratio of bi-BAs allotropes

	AA(B-As)	AB(B-As)	AA(B-B)	AB(B-B)	AA(As-As)	AB(As-As)	h-BAs
v_{TA} (km/s)	5.07	5.03	5.12	4.97	5.10	5.07	5.54
v_{LA} (km/s)	7.84	7.79	7.83	7.64	7.81	7.84	9.42
C_{11}-C_{22} (N/m)	177	172	179	169	175	172	126
C_{12} (N/m)	30	29	28	28	31	30	35
C_{66} (N/m)	74	71	75	70	72	71	47
Y (N/m)	172	167	175	164	170	167	116
Poisson ratio	0.169	0.169	0.156	0.166	0.177	0.174	0.278

Table 15. Lattice thermal conductivity, lattice thermal conductance and mean free path of bi-BAs allotropes and h-BAs at 300K

	AA(B-As)	AB(B-As)	AA(B-B)	AB(B-B)	AA(As-As)	AB(As-As)	h-BAs
κ (W/mK)	176	174	209	197	313	230	182
C (GW/m²K)	0.82	0.77	0.83	0.81	0.79	0.79	1.06
Λ (nm)	226.0	214.6	251.8	243.2	396.2	291.1	171.7

Machine learning and molecular dynamics methods were combined[63] in order to predict the thermal conductivity and assess the effect of anharmonicity upon the thermal transport properties. The machine-learning method was based upon a matrix tensor algorithm which could accurately describe the lattice dynamics of BAs. A phonon spectral energy density analysis showed that the machine-learning technique could model both phonon-mode softening and a linewidth-broadening which was induced by anharmonicity at finite temperatures. Based upon the phonon Boltzmann transport equation and the 3-phonon scattering process, the calculated results showed that the accuracy of machine learning in predicting thermal conductivity was comparable to that of density-functional theory. The machine learning method greatly overestimated the conductivity when compared with experimental results, due to the marked influence of high-order phonon scattering processes. The conductivity values which were predicted by equilibrium molecular dynamics simulations, when combined machine learning, agreed well with experimental data.

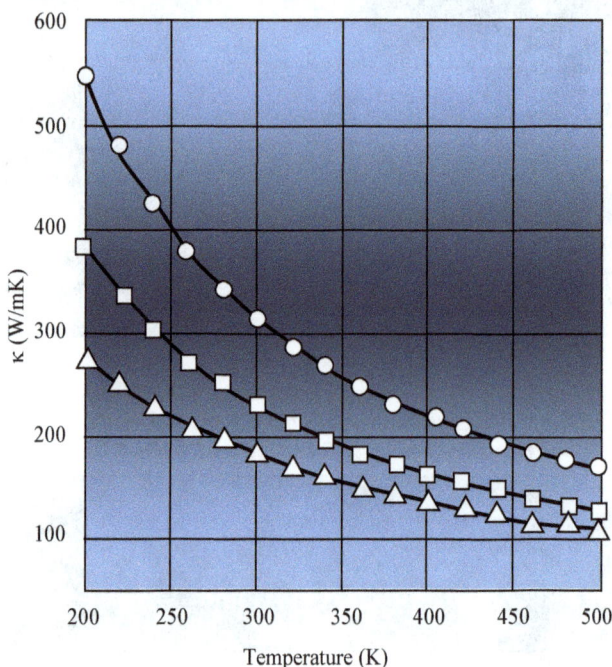

Figure 2. Lattice thermal conductivity of AA(As-As), AB(As-As) allotropes and h-Bas Circles: AB(As-As), squares: AA(As-As), triangles: h-BAs

Thermal phononics theory tends to rely on the wave-interference of phonons and is fully efficacious only at sub-Kelvin temperatures or at atomic scales. An alternative approach has been proposed[64] which is based upon a particle model of phonons and their ballistic transport. This approach in effect treats ray phononics as a thermal analogy of ray optics. The former is free from the limitations of traditional phononics and permits the creation and guidance of thermal fluxes in realistic nanostructures; regardless of their surface roughness. Simulations have demonstrated some possible applications of ray phononics to the directional emission of heat rays, to the filtering of phonon spectra and to the shielding of a chosen region from a thermal gradient. The method is not limited to low temperatures, and ray-phononic nanostructures can potentially control heat fluxes even at room temperature.

A comprehensive study was made[65] of the thermal diffusive properties of 2-dimensional monolayer honeycomb BAs by solving the phonon Boltzmann transport equation using first-principles calculations. These showed that the high thermal conductivity (181W/mK) at 300K was mainly due to in-plane phonon modes rather than to the ZA mode. This result was explained by the existence of an unique frequency-independent so-called platform region in the relaxation-time of the in-plane phonons. On the basis of the selection rule, it was found that this platform arose from a suppressed ZA + ZA → IA scattering channel. The IA modes made large contributions to the thermal conductivity, unlike the cases of graphene and hexagonal BN. This was because the IA modes not only had long relaxation times and high group-velocities, but also a flat ZA branch due to a weak π-bond strength. A comparative study was made of the thermal conductivity of 2-dimensional hexagonal BAs and hexagonal GaN because both materials have a similar density. The thermal conductivity of hexagonal BAs was an order-of-magnitude higher than that (16W/mK) of hexagonal GaN, which is governed by a different phonon-scattering process which is attributed to the opposite wave-vector dependence of out-of-plane optical phonons. Upon comparing the phonon dispersion of hexagonal BAs with hexagonal GaN, the main difference was found to be that the ZO branch of the latter exhibited an obvious downward trend in going from the BZ center to the boundary, which increased the phase space of ZA + IA → ZO scattering; leading to the low thermal conductivity. With regard to long-range electrostatic interactions it was found that the LO (longitudinal optical)–TO (transverse optical) splitting appeared at the Brillouin zone center and resulted in a small frequency-gap (0.31THz). This slight shift of the LO branch arose from the similar electronegativities of the boron and arsenic atoms (table16). The main differences in phonon spectrum of hexagonal BAs and the majority of other planar hexagonal III–V compounds was the flat ZA branch and upward ZO branch. In hexagonal BAs the maximum frequency of the ZA branch was only 1.81THz; thus indicating that ZA phonons indeed had a low group-velocity and that the out-of-plane vibration was very weak. This could be explained by the large (85.73g/mol) atomic density of hexagonal BAs as compared to that (24.81g/mol) of hexagonal BN, and by the weak B–As bonding along the out-of-plane direction (table 17).

Table 16. Born effective charges, Z, of boron and arsenic
atoms and dielectric constants, ε, of hexagonal BAs

Component	Z (B)	Z (As)	ε
xx	1.744	-1.744	4.678
yy	1.743	-1.743	4.678
zz	-0.051	0.051	1.162

Table 17. Anharmonic force constants of bonding atoms in graphene, hexagonal-BN and
hexagonal BAs, where x corresponds to the longitudinal direction of the bond

Anharmonic Force Constant	C-C (eV/Å^2)	B-N (eV/Å^2)	B-As (eV/Å^2)
Φ_{xx}	25.21	21.72	12.70
Φ_{yy}	10.47	8.05	2.65
Φ_{zz}	6.11	5.03	0.54

Isovalent boron incorporation into GaAs without phase separation was possible[66] under high-arsenic and reduced-boron exposure, and a BAs phase could form under more boron-rich conditions. More gallium-rich conditions led to boron substitution at arsenic positions and the boron dimers which formed could be the starting-point for the formation of a boron phase whereas true antisite boron substitution was less probable.

The efficient synthesis of high-quality BAs crystals has limited their use. An early attempt to prepare boron-rich compounds by using a nickel flux method produced only the boron sub-arsenide, $B_{12}As_2$. Thermodynamic considerations suggest that $B_{12}As_2$ is stable at temperatures greater than 920C. Below 920C, sub-stable BAs compounds could be obtained only by using high-temperature synthesis methods involving gaseous arsenic and solid boron[67]. A convenient means for improving the quality and yield of single crystals is to use a crystal seed. Chemical vapor transport is a promising method for producing high-quality BAs crystals and seeds. But this method requires times of the order of weeks or months, plus careful seed selection. Available synthesis routes all involve low synthesis rates, polycrystalline particles and defective structures which are attributed to the fluctuation of components during mass transfer. The use of transport agents greatly improves the efficiency of reaction and growth, and commonly used agents such as I_2 permit the transport efficiency of a boron source to attain some 90%. Other effective agents include AsI_3 TeI_4 and NH_4I. Elemental iodine halogenates the boron source so as to produce gaseous B–I compounds. This change converts a transfer process from solid-state diffusion to gaseous convection. Defects in the product, such as twins, vacancies, interstitial atoms and antisites cause deviations from the theoretical stoichiometric ratio.

An investigation was made of the effect of iodine-containing transport agents, I_2 and BI_3, upon BAs which is grown using chemical vapor transport. Like I_2, the BI_3 accelerated synthesis and improved the mass-fraction of BAs from 12% to over 90% at 820C under a pressure of 1.5MPa. Such improvements were beyond the promotional effects of increased temperature and pressure. Both agents improved the quality of BAs crystals by reducing the full-width at half-maximum by up to 20%. The I_2 agglomerated grown crystals with twins (50nm wide) while BI_3 improved the crystal anisotropy and uniformity of BAs crystals with narrow twins (15nm wide) and increased the stoichiometry ratio from 0.990 to unity. Due to boron interstitials from the excess boron, the spacing of layers on $\{111\}$ increased to 0.286nm in the presence of I_2. Due to its coordinated effect, the BI_3 affected only slightly influenced the layer spacing at 0.275nm; close to the theoretical value of 0.276nm. During chemical vapor transport, anisotropic crystals with flat surfaces exhibited single-crystal characteristics under the influence of BI_3. Unlike I_2, the coordinated effect of BI_3 could promote the efficient preparation of high-quality crystal seeds and advance the exploitation of BAs.

In addition to accelerating reactions, the transport agents, especially BI_3, can change the crystal characteristics. A preparation which was conducted at 820C under a pressure of 1.5MPa showed that transport agents changed the growth and features of BAs crystals. Reaction without agents dispersed a few small crystals with a size of about 5μm on coral-shaped boron powders. These crystals clustered together like grapes. When BI_3 or I_2 is added, the crystals become the predominant phase. Unreacted B powder is found around BAs crystals and are not detected by X-ray diffraction, due to their low content. Crystals cluster around like apples under the influence of I_2 and agglomerate into spherical polycrystals of over 50μm in size. Many boundaries appear on the crystal surface. When BI_3 is added, the growth of individual polyhedral crystals can be expected because of clear facets between individual crystals. The differing transport effects of the agents upon elemental boron are reflected by the product. The average atomic ratio of boron and arsenic in BAs crystals is 55.54:41.83 (when the stoichiometric ratio is 1.328) in one case and is 49.28:50.16 (ratio of 0.982) in another case (table 18). Impurities such as Si and I are also detected and the contents may also differ (table 19). A large crystal size also reflects the marked effect of the iodides.

Table 18. Composition of elements in samples with differing elemental ratios

Element	Content (%)	B/As Ratio
B	55.54	1.328
Si	0.13	1.328
I	2.50	1.328
As	41.83	1.328
B	49.28	0.982
Si	0.46	0.982
I	00.10	0.982
As	50.16	0.982

Small cubic crystals of boron arsenide were later produced[68] via chemical vapour transport, using halogens as a transport reagent. Boron arsenide which was produced from the elements was transported through a 850-to-400C temperature gradient using 1.6atm of iodine as the transport agent. It was suggested that the reaction of iodine with boron arsenide involved the formation and transport of a boron sub-iodide, with the arsenic subliming directly to the cooler end of the tube. The transport of elemental boron was used to produce the alpha-rhombohedral form of BAs by starting with 4N5 boron, and iodine. By assuming that the transporting species was boron tri-iodide, calculations could be made to determine the conditions for transport. Upon using the calculated transport conditions of 900 to 1100C and a pressure of 1.6atm of iodine, no appreciable material was produced. When the temperature gradient was lowered and reversed (900-400C), at the same pressure, boron was transported to the cooler end. When using iodine, the transported boron was amorphous. In order to ensure a crystalline product, bromine was substituted for iodine on the basis that bromine would form a more stable intermediate and thus allow transport to a higher minimum temperature where crystal growth would be feasible. When using bromine, deposition occurred in the 500 to 600C range. The product was found to consist of beta-rhombohedral boron, plus some beta-tetragonal boron.

In another case, boron arsenide which had been produced by combining the elements was transported through a temperature gradient of from 850 to 400C by using a 1.6atm of iodine as the transporting agent. The product consisted of mainly boron arsenide. The appearance of beta-rhombohedral boron was unexpected given that experience had shown that alpha-rhombohedral boron could be prepared epitaxially on orientated silicon and fused silica substrates at up to 1050C. A temperature of 1000C was generally required in order to convert amorphous boron into beta-rhombohedral boron, but here it seemed that the beta-rhombohedral form was directly deposited in crystalline form, and not as an amorphous material although the deposition temperature was considerably less than 1000C.

The similarity of the reaction conditions which obtained when using iodine as the reagent suggested that boron and boron-group vapour transport occurred via the same intermediate, this in turn implied that the reaction of iodine with boron arsenide involved the formation and transport of a boron sub-iodide, with the arsenic subliming directly to the cooler end of the tube. This was confirmed by the appearance of arsenic which was mixed with the transported boron arsenide. Overall, small crystals of boron and boron arsenide were produced by chemical vapor transport down a temperature gradient.

The equilibria existing during the chemical transport of arsenides of the third group with iodine were studied[69] using Raman spectroscopy of the gaseous phases at up to 900C. At above 300C, AlAs, GaAs and InAs reacted fully and irreversibly with iodine to give a metallic tri-iodide. Mono-iodides then formed at above 750C. The results generally supported the supposed mechanism, but boron arsenide reacted with iodine near to 350C to give arsenic tri-iodide without any boron tri-iodide. At higher temperatures, chemical attack of the silica container occurred and SiI_4 slowly replaced AsI_3. Because of partial dissociation of AsI_3 and SiI_4, transport could be explained by invoking additional equilibria:

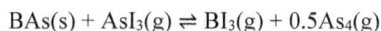

$$BAs(s) + AsI_3(g) \rightleftharpoons BI_3(g) + 0.5As_4(g)$$

and

$$BAs(s) + 1.5I_2(g) \rightleftharpoons BI_3(g) + 0.25As_4(g)$$

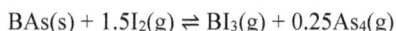

A much more recent investigation was made[70] of the effect of iodine-containing transport agents such as I_2 and boron tri-iodide upon BAs which was prepared by chemical vapour transport. The presence of BI_3 accelerated the synthesis and increased the mass-fraction of BAs from some 12% to more than 90% at 820C and 1.5MPa, after factoring-out the enhancing effect of increased temperature and pressure alone. Both agents improved the quality of BAs crystals by reducing the full-width at half-maximum by 10 to 20%. The presence of I_2 produced crystals with twin defects that were about 50nm wide, while BI_3 improved the crystal anisotropy and compositional uniformity of BAs crystals with narrow twins (~15nm wide) and increased the stoichiometry-ratio almost to unity. Due to

the boron interstitials arising from the excess boron supply, the spacing of layers on 111}
increased to 0.286nm in the presence of I_2. Meanwhile BI_3 increased the layer-spacing
only slightly, to 0.275nm; close to the theoretical value of 0.276nm. Under the action of
BI_3 during chemical vapour transport, anisotropic crystals with flat surfaces exhibited
monocrystalline characteristics and, unlike I_2, the tri-iodide could promote the efficient
preparation of high-quality BAs crystal seeds. Single crystals of cubic BAs, of up to a
few mm in size, were grown[71] via a chemical vapour transport method which was based
upon a TeI_4 transport agent and involved gas pressures of up to 8atm. Raman
spectroscopy revealed a sharp P1 phonon mode, indicating a good crystalline quality.
High-angle annular dark-field scanning transmission electron microscopy revealed the
antisite pairs, As_B (As atom on B site) and B_{As} (B atom on As site). A bulk thermal
conductivity of 133W/mK was measured at 300K by performing steady-state
comparative measurements.

In one of the first theoretical studies[72] of the material, first-principles self-consistent
orthogonalized-plane-wave energy-band calculations were carried out for cubic BAs by
using a non-relativistic formalism and Slater's free-electron exchange approximation. The
imaginary part of the dielectric constant, spin-orbit splittings, effective masses,
deformation energies and X-ray form-factors were predicted. The model used did not
have any adjustable parameters, but it was necessary to impose the lattice constant. The
above value of 4.777Å was used although a value of 4.7778Å had been reported. In any
case, values of 4.767 and 4.787Å were also assumed in order to determine the effect of
pressure. It was found that there were 2 minima in the bottom conduction band, both of
which were lower than the minimum at Γ. The lowest minimum occurred at 0.81 of the
distance from the Γ-point to the x point. The indirect gap which was measured to this
point was 1.6eV. The next lowest minimum occurred at the L-point, with a figure of
2.93eV. The direct gap was 3.56eV. The fundamental electronic and optical properties
were again determined[73] by using density functional theory and many-body perturbation
theory, including quasi-particle and spin-orbit coupling corrections. The fundamental
band-gap was indirect, with a value of 2.049eV, while the minimum direct gap was
4.135eV. The calculated carrier effective masses were smaller for holes than for
electrons, thus indicating a higher hole mobility and easier p-type doping. A small
difference between the static and high-frequency dielectric constants indicated that BAs
is only weakly ionic. The imaginary part of the dielectric function included a marked
absorption peak which corresponded to parallel bands in the band structure. An estimated
exciton binding-energy of 43meV indicated that the excitons were relatively resistant to
thermal dissociation at room temperature.

A first-principles calculation was made of the energy bands of BAs, with no adjustment being applied in order to fit experiment. The only factor used was the lattice constant. The choice of the crystal potential included Slater exchange, on the grounds that it gives a better agreement with experiment than do other forms of exchange. This is suggested to be because correlation is effectively incorporated within the Slater exchange. The composite wave variational version of the augmented plane wave method was used[74] to calculate the electronic band structure of boron arsenide at the high-symmetry points, Γ, X and L. The tight-binding interpolation scheme of Slater and Koster was used to deduce the remainder of the band structure. In this early work, BAs was found to be an indirect-gap material with a valence-band maximum at Γ_{15v} while the conduction-band minimum was at X_{3c}. The Γ_{1c} level was higher than the Γ_{15c} level. The calculated value of the Γ_{15v}-X_{3c} indirect gap was 1.27eV, as compared with an experimental value[75,76] of 1.4eV.

The overlap integrals were found simply by interpolation parameters, rather than being deduced from physical considerations. They were consistent with the fact that ppσ > ppπ because the pσ wave-functions overlap less than do the pπ wave-functions. There was however some doubt concerning the non-uniqueness of these interpolation parameters. The set of parameters which corresponded to the best fitting was chosen. The maximum in the imaginary part of the dielectric constant, as a function of photon energy, was at 0.305Ryd due to the Γ_{3v}-Γ_{3c} transition, followed by a shoulder at 0.311Ryd due to the Γ_{15v}-Γ_{15c} transition. The second peak was at 0.405Ryd, due to the transition, X_{5v}-X_{3c}. An experimental peak had been measured at. 0.351Ryd.

A full potential linearized augmented plane wave method was used[77] to predict the exchange-correlation potential, ground-state properties, lattice parameter, bulk modulus and pressure-derivative. The valence charge density for the equilibrium lattice constant revealed an inverse relationship between cations and anions.

The electronic and optical properties of the zincblende and rock-salt phases were studied[78] by using density functional theory in the generalized gradient approximation. Using enthalpy-pressure data, a structural phase transition from zincblende to rock-salt was observed at 141GPa. Calculated electronic properties showed that zincblende BAs is a semiconductor, while rock-salt BAs is a semi-metal.

An *ab initio* self-consistent density functional theory study was made[79] of the electronic, transport and bulk properties of zincblende boron arsenide. The Ceperley-Alder local density approximation potential was used, together with the linear combination of Gaussian orbitals formalism and the Bagayoko-Zhao-Williams method to perform entirely self-consistent calculations. These results had the full physical content of density functional theory and included electronic energy bands, densities of states, effective masses and the bulk modulus. The predicted indirect band-gap of 1.48eV, from Γ to a conduction band minimum close to X, for a room-temperature lattice constant of 4.777Å, was in excellent agreement with the experimental value of 1.46eV.

The electronic band structure and optical properties were also explored by using[80] a relativistic quasi-particle self-consistent approach which included electron-hole interactions via solution of the Bethe-Salpeter equation. The electronic and optical properties were also studied using standard and hybrid density functional theory. The inclusion of self-consistency and vertex corrections led to a marked improvement in the calculated band features. Comparison of the results to previous calculations which had been performed by using the single-shot GW approach or density functional theory methods, revealed an appreciable scatter in the calculated indirect and direct band-gaps. It was predicted that BAs has an indirect gap of 1.674eV and a direct gap of 3.990eV; values which were consistent with experimental data and similar theoretical predictions. Hybrid density functional theory predicted the indirect accurately, but gave less accurate predictions for other band features such as spin-orbit splittings. The values found for the Born effective charges and dielectric constants confirmed the anomalous covalent bonding characteristics of the material.

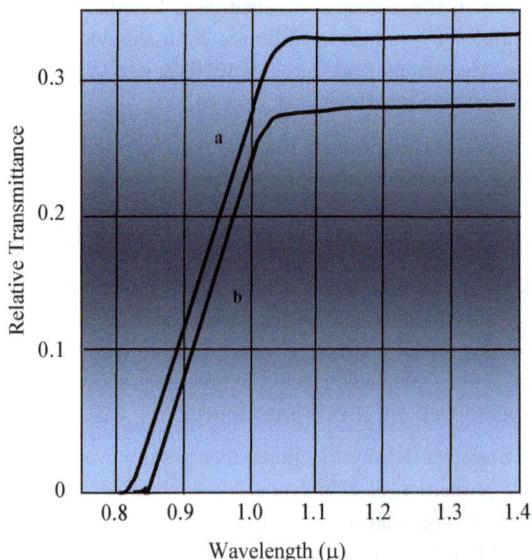

Figure 3. Optical transmission at 300K of cubic and rhombohedral boron arsenide
a: rhombohedral boron-sub-arsenide, b: cubic boron arsenide

By using optical transmission methods and powdered samples the energy-gaps of the cubic and rhombohedral forms were estimated[81] to be 1.46 and 1.51eV, respectively. Mixed crystals of boron and gallium arsenides were also prepared via vapour-transport. The energy-gap varied from 1.36 to 1.27eV as the boron content ranged from 0.8 to 2.8wt%. Measurements of the effects of temperature and injection-level upon the emission of diodes which were prepared from the mixed crystals revealed that all of the compositions with up to 2.8wt%B were direct semiconductors.

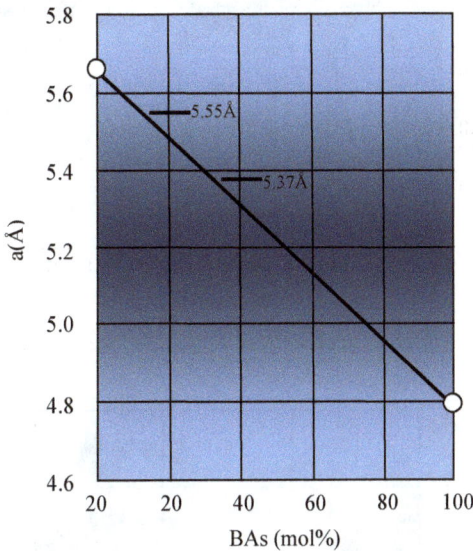

Figure 4. Lattice constant as a function of the composition of $B_xGa_{1-x}As$

The Hall mobilities and free-carrier concentrations were calculated (table 19), showing that a much lower Hall mobility was observed for the mixed crystals than for gallium arsenide having a comparable free-carrier concentrations. In 0.78wt%B samples with a free-carrier concentration of 1.6 x 10^{18}/cm^3 the Hall mobility was some 5 times lower than that of GaAs with the same free-carrier concentration. The Hall mobility of the 2.8wt%B composition was roughly halved when the free-carrier concentration was about half that of the lower boron-content sample.

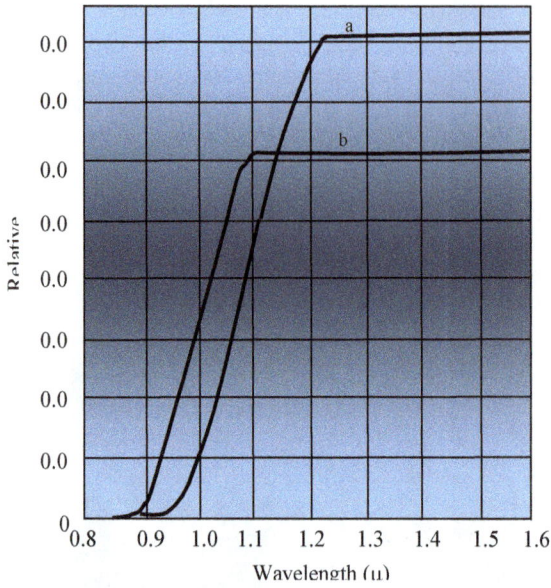

Figure 5. Optical transmission at 300K of $B_xGa_{1-x}As$ for different thicknesses
a: 3mil, b:5.5mil

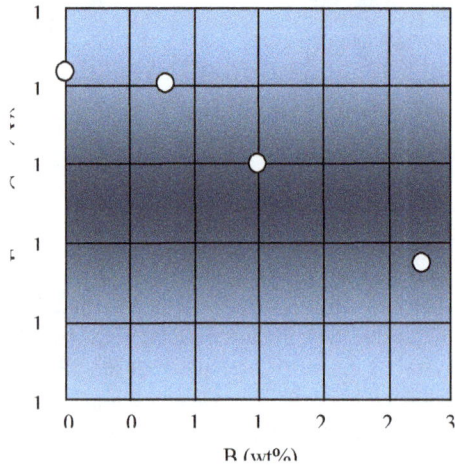

Figure 6. Emission energy-gap at 300K of $B_xGa_{1-x}As$ for various boron contents

Table 19. Electrical properties of $B_xGa_{1-x}As$ mixed crystals

B(wt%)	Carrier Concentration(/cm^3)	Hall Mobility (cm^2/Vs)	ρ(Ωcm)
0.78	1.6 x 10^{18}	515	0.0073
2.80	7.3 x 10^{17}	280	0.031

In a further refinement[82] of the preparation methods, single crystals of boron arsenide which were several mm in size were produced by the chemical transport of polycrystalline boron arsenide in a temperature gradient. The transported crystals were p-type and had a resistivity of about 0.01Ωcm at temperatures of 77 to 500K. The hole concentration and the Hall mobility were almost independent of the temperature at 77 to 300K, and were of the order of 10^{18} to 10^{19}/cm^3 and 100 to 400cm^2/Vs, respectively.

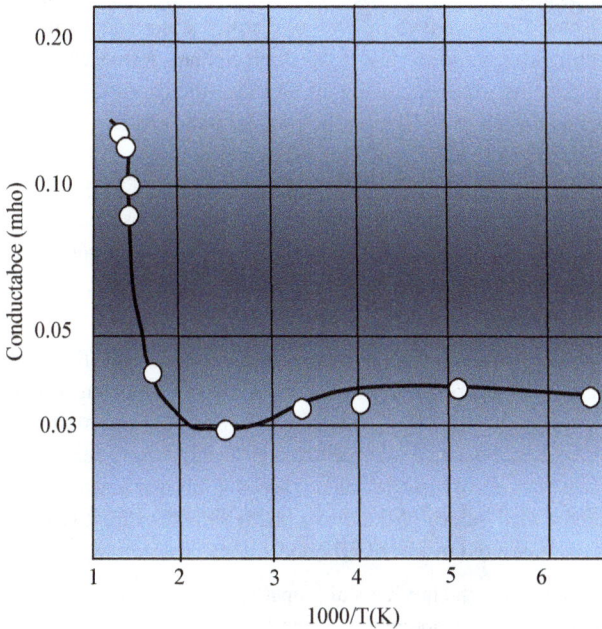

Figure 7. Electrical conductance of boron arsenide crystal as a function of temperature

Cubic boron arsenide films could be deposited[83] onto the basal plane of hexagonal silicon carbide, {111}-oriented sodium fluoride and silicon substrates at 800 to 850C via the thermal decomposition of a diborane-arsine mixture in a hydrogen atmosphere. The films which were deposited onto silicon carbide were randomly oriented and adhered to the substrate. The boron arsenide deposits on sodium fluoride substrates did not adhere, due to the large differences in the thermal expansion coefficients. The arsenide films which were deposited onto silicon substrates were amorphous. Optical absorption measurements again indicated that boron arsenide was a direct-gap material with a room-temperature energy-gap of about 1.45eV. The current-voltage characteristics of arsenide-silicon structures suggested that the current-controlling mechanism was similar to that of an insulator which contained traps of uniform energy, and the density of traps was estimated to be about $10^{17}/cm^3eV$.

A wet chemical route was proposed[84] for the preparation of BAs. Nanoparticles were prepared by using a hydrothermal method, followed by sintering in an argon atmosphere. Precursors of As_2O_3 and $NaBH_4$ were used in both stoichiometric and non-stoichiometric ratios. The As_2O_3 was first dissolved in 40ml of distilled water and a solution of $NaBH_4$ was later added drop-wise while stirring. A black precipitates formed. Broad peaks in the X-ray diffraction patterns were tentatively attributed to the nano-size or amorphousness of the sample. A broad peak in stoichiometric samples indicated an incomplete reduction of As_2O_3, as this peak was absent from non-stoichiometric samples due to excess $NaBH_4$. Two other broad peaks closely matched BAs. The formation of BAs was confirmed on the basis of the interatomic spacing. Crystalline BAs was thus successfully synthesized by using a simple wet chemical route, using arsenic trioxide and sodium borohydride. The yield however was very low and it is possible that boron escaped by forming diborane gas.

The complex dielectric function, refractive index and absorption coefficient of BAs in the ultra-violet, visible and near-infrared wavelength ranges were measured[85] at room temperature by using spectroscopic ellipsometry, plus transmission and reflection spectroscopy. The optical response was predicted by using density functional theory and many-body perturbation theory while taking account of quasi-particle and excitonic corrections. The predicted values of the direct (4.25eV) and indirect (2.07eV) band-gap agreed well with the measured results of 4.12 and 2.02eV, respectively.

Also of potential interest are the mechanical properties of boron arsenide. Two theoretical methods were applied[86] to the prediction of the bulk moduli of solids. One was based upon an *ab initio* concept which required knowledge only of the atomic numbers of the constituent atoms. The other was a simpler empirical approach which was applicable to diamond- and zincblende-structured solids and used the nearest-neighbour bond-length as input (table 20). Both methods gave comparable results, but the first-principles approach could clarify the details of the structural, bonding and electronic properties. For III-V compounds which are formed from elements which are not in the first row of the periodic table, a typical charge-density plot for the valence electrons shows a partially covalent partially ionic bond configuration. The ionic component occurs because the anion accumulates more valence electron charge because of its more positive core potential. In the case of BAs a covalent-like charge distribution is also found with a slightly higher charge density near to the boron sites. This unusual situation is the reverse of what is found in other III-V compounds, and boron may be more accurately described as being an anion rather than a cation. The strong ionic-covalent bonding in BN results in a short bond-length and high bulk modulus, while the weaker bonds of BP and BAs yield longer bond-lengths and lower bulk moduli; even though both are highly covalent systems. The predicted value of the bulk modulus of BAs ranged from 138 to 145GPa.

Table 20. Lattice constants and bulk moduli of BAs and other materials

Material	Method	a (nm)	B (GPa)
BAs	calculated	04777	145
AlAs	experimental	0.243	77.0
AlAs	calculated	0.243	78.3
AlP	experimental	0.236	86.0
AlP	calculated	0.236	86.7
AlSb	experimental	0.266	58.2
AlSb	calculated	0.266	57.0
BN	experiment	0.3615	465
BN	calculated	0.3606	367
BP	experiment	0.4538	173
BP	calculated	0.4558	165
C	experiment	0.3567	443
C	calculated	0.3561	438
CdS	calculated	0.252	62.0
CdS	experimental	0.252	60.3
CdSe	calculated	0.262	53.0

CdSe	experimental	0.262	52.6
CdTe	calculated	0.281	42.4
CdTe	experimental	0.281	41.2
GaAs	experimental	0.245	74.8
GaAs	calculated	0.245	76.1
GaP	experimental	0.236	88.7
GaP	calculated	0.236	86.7
GaSb	experimental	0.265	57.0
GaSb	calculated	0.265	57.8
HgSe	calculated	0.263	50.0
HgSe	experimental	0.263	51.9
HgTe	calculated	0.278	42.3
HgTe	experimental	0.278	42.7
InAs	experimental	0.261	60.0
InAs	calculated	0.261	61.0
InP	experimental	0.254	71.0
InP	calculated	0.254	67.0
InSb	calculated	0.281	47.4
InSb	experimental	0.281	47.1
Si	experiment	0.5429	99
Si	calculated	0.5433	92
SiC	experiment	0.4360	224
SiC	calculated	0.4361	212
ZnS	calculated	0.234	77.1
ZnS	experimental	0.234	78.1
ZnSe	calculated	0.245	62.4
ZnSe	experimental	0.245	66.5
ZnTe	calculated	0.264	51.0
ZnTe	experimental	0.264	51.2

Monolayers

The bulk semiconductors, BAs and AlN, have indirect and direct gaps, respectively, on the other hand, electronic calculations demonstrate that one layer, or a few layers, of BAs and AlN can exhibit a graphite-like structure with other electronic properties. Infinite-sheet single-layer heterojunction structures, based upon alternating strips of honeycomb BAs and AlN layers were investigated[87] by using first-principles density functional theory calculations. The optimum geometries, density-of-states, band-gaps, formation energies and wave functions were determined for various strip-widths which were joined along zig-zag and armchair edges. In optimum heterojunction geometries, the BAs narrow strips had a corrugated form, due to lattice mismatch. Zig-zag heterojunctions were more energetically favoured than were armchair heterojunctions. The formation energy went through a maximum at the point where the heterojunction became a planar structure. Electronic charge-density results indicated a more ionic behaviour in the case of Al-N bonds than in the case of B-As bonds; in agreement with monolayer results. The conduction-band minimum for both heterojunctions involved confined states that were located mainly at AlN strips. The valence-band maximum involved confined states which were located mainly at BAs strips.

First-principles calculations were made[88] of the ideal strengths and band-alignments of polar BAs/GaN heterojunctions and superlattices (tables 21 and 22). This showed that, under normal compression conditions, all BAs/GaN interface configurations exhibited a much higher compressive stiffness when compared to that in the bulk GaN [00•1] direction., with its softening during its structural transformation under compression being greatly suppressed. This helped to improve protection of the electronic properties under impact. The natural band alignments of the mismatched BAs/GaN heterojunctions are calculated, and most heterojunction and all superlattice interfaces exhibited type-II staggered band offsets. Large built-in polarization electric fields were present in superlattices, with repeated positively and negatively charged N-As and Ga-B interfaces. This produced a saw-tooth dipole potential, which could distribute electrons and holes to different interfaces; again relevant to photocatalysis. The BAs/GaN heterojunction could constitute an alternative to GaN-on-diamond heat dissipation systems and also had possible application in photovoltaic devices as a type-II semiconductor heterojunction or superlattice.

Table 21. Lattice parameters of orthorhombic super-cell, lattice mismatch, Heyd–Scuseria–Ernzerhof band-gaps, valence-band maximum positions relative to the macroscopic-averaged electrostatic potential of bulk BAs and GaN and the deformation potential energy in each direction

Parameter	BAs	GaN
a_1(Å)	3.406	3.219
a_2(Å)	5.899	5.375
a_1(%)	-5.50	0
a_2(%)	-5.50	0
E_g(eV)	1.90	3.51
E_v(eV)	5.25	3.84
$\Delta E_D[a_1]$(eV)	-0.73	0
$\Delta E_D[a_2]$(eV)	-0.72	0

Table 22. Dipole potential energy, valence-band offset and conduction-band offset, with and without deformation potential energy of BAs/GaN heterojunctions for various interface configurations

Parameter	B-N bonding	B-Ga bonding	As-Ga bonding	S-N bonding
ΔV_P(eV)	-0.52	1.52	0.15	0.26
VBO without E_D(eV)	-1.93	0.11	-1.26	-1.15
VBO with E_D(eV)	-3.38	-1.34	2.71	-2.60
CBO without E_D(eV)	-0.32	-1.72	0.38	0.46
CBO with E_D(eV)	-1.77	0.27	-1.10	-0.99

Boron arsenide also possesses a phonon density-of-states which overlaps perfectly with those of GaN and metal layers between GaN and BAs. This results in an eightfold reduction in thermal boundary resistance in a GaN-on-BAs system, as compared to that in a GaN-on-diamond system. Having an indirect band-gap structure and a high optical absorption coefficient, it can also operate as an active electronic and opto-electronic component in addition to offering cooling. Boron arsenide has both a high hole and electron mobility, 2110 and 1400cm^2/Vs, and native p-type dopability together with a low mass density, low effective carrier mass and chemical stability at ambient temperatures. The lattice constant of 3.38Å of BAs on its (111) plane and that, 3.19Å, of GaN on its (00•1) plane, leads to a mismatch of 5.6% due to formation of the BAs/GaN heterojunction. The ideal tensile and compressive strengths and band alignments for BAs(111)/GaN(00•1) heterojunctions and superlattices with 4 interface configurations were investigated, and showed that the B-N and B-Ga interface configurations were the most stable heterojunctions, with their normal tensile strengths being nearly as high as those in bulk BAs[111] and GaN[00•1]. The As-Ga and As-N interfaces are much weaker and break up under normal tensile stresses. Under normal compression, all of the BAs/GaN heterojunction interfaces had a higher compressive stiffness, as compared to that in the bulk GaN[00•1] direction, where the zincblende BAs suppresses the softening of wurtzite GaN during a structural transformation from P63mc to P63/mmc symmetry under compression stresses. The high compressive stiffness of the BAs/GaN heterojunction might protect electronic properties from impact. The natural band alignments of the mismatched BAs/GaN heterojunctions were calculated, and this showed that most of the interfaces gave rise to type-II staggered systems. Only the B-Ga interface exhibited type-I straddling band alignment. In BAs/GaN superlattices all of the natural band alignments of heterojunction interfaces were type-II. In superlattices with repeated positively- and negatively-charged N-As and Ga-B interfaces, a large polarization built-in electric field existed and produced a sawtooth polar dipole potential which could effectively confine electrons and holes to different interfaces; something desirable for photocatalytic purposes. Calculations showed that BAs/GaN heterojunctions could be an alternative to GaN-on-SiC or GaN-on-diamond heat dissipation systems and also offer new applications in photovoltaic, photodiode and photocatalytic devices.

The flexibility, tunability and active surfaces of ultra-thin films were explored[89], using time-dependent density functional theory, for configurations having Archimedean (4,8)-tessellations which exhibited exceptional light and matter interactions. Planar monolayers of haeckelite boron-pnictogen binary materials such as BAs exhibited strong interband absorbance in the ultra-violet region (boron-nitride nanosheet) and in the infra-red region (heavier pnictogens); thus offering a means for the photo-capture of high-density solar photons. Buckled haeckelite boron arsenide exhibited a similar, but slightly muted, optical response that was blue-shifted from its planar configuration. The marked optical response of the ultra-thin films was attributed to the unique band-structure localization of π-electrons in the ground state, Van Hove singularities at band-extrema and complementary elemental properties. This (4,8) haeckelite motif showed that many 2-

dimensional films having very different lattice tessellations to those of known ultra-thin materials could have potential value.

First-principles density functional theory calculations were used[90] to study the stability and properties of hydrogenated and fluorinated 2-dimensional sp^3 boron phosphide and boron arsenide. The phonon dispersion spectrum and phonon density-of-states of the hydrogenated and fluorinated BX systems were found to be different, and this was attributed to the differing masses of hydrogen and fluorine. The hydrogenated systems had larger, indirect, band-gaps and were different to fluorinated BX systems. They could be used in hydrogen-storage applications and ultra-fast electronic devices. Unlike the electronic properties determined of fluorinated/hydrogenated monolayers, the electronic properties of bilayers could be finely tuned to impart a narrow-gap semiconducting, metallic or nearly semimetallic nature by choosing a suitable arrangement of layers. A nearly-linear dispersion of the conduction band-edge and heavy-hole and light-hole bands were interesting characteristics. Exceptional values of the effective masses guaranteed fast electronic transport.

The electronic and phonon transport properties of graphene-like boron phosphide, boron arsenide and boron antimonide monolayers were investigated[91] (tables 23 and 24) using first-principles calculations combined with Boltzmann theory. By taking account of both phonon-phonon and electron-phonon scattering, it was demonstrated that strong bond-anharmonicity in the BAs and BSb monolayers could markedly suppress the phonon relaxation time but barely affected that of electrons. Both systems therefore exhibited comparable power-factors to that of BP monolayers, but much lower lattice thermal conductivities. A maximum p-type figure-of-merit value above 3.0 could be obtained in both BAs and BSb monolayers for optimum carrier concentrations. A very similar p-type and n-type thermo-electric behaviour was observed in BSb monolayers along the zig-zag direction.

Table 23. Comparison of acoustic phonon group velocities (300K) of BX monolayers

Mode	BP	BAs	BSb
ZA (km/s)	3.2	1.6	1.4
TA (km/s)	7.9	6.2	4.1
LA (km/s)	13.9	10.1	7.5

Table 24. Room-temperature electronic transport coefficients of p-type BX monolayers
with optimized carrier concentration along the armchair direction

Property	BP	BAs	BSb
n (10^{19}/cm^3)	4.9	1.6	1.4
S (μV/K)	199	240	251
σ (10^4S/cm)	2.28	3.16	3.29
κ (W/mK)	5.97	8.28	6.53
S$^2\sigma$ (W/mK^2)	0.09	0.18	0.21
μ (m^2/VK)	0.29	1.20	1.66

Two-dimensional BAs monolayers possess useful properties which arise from their high thermal conductivity and bipolar magnetic properties, thus making them suitable candidates for spintronic devices. Such 2-dimensional BAs-based electronic devices unfortunately degrade when exposed to air. The stability of 2-dimensional BAs-based systems was studied[92] with regard to their interaction with oxygen molecules. Calculations which were based upon density functional theory suggested that the O_2 molecule could spontaneously dissociate on the surface of a monolayer BAs system. During this dissociation process it underwent an exothermic reaction, with an evolved energy of 2.60eV. The electronic band-gap, without and with the O_2 molecule, was 0.77 and 0.68eV, respectively, with a direct band-gap. Optical studies showed that, in pure BAs, the higher absorption of photon energy was in the ultra-violet region but, in oxygen-adsorbed BAs, the higher absorption of photon energy shifted into the infra-red as well as the visible region. The oxidized surface of monolayer BAs-based electronic devices was accordingly slightly degraded in air.

The doping effect of transition metals (vanadium, chromium, manganese) upon the structural, electronic and magnetic properties of boron arsenide monolayers was investigated[93] using pseudopotential methods within the framework of density functional theory.

The first principles pseudopotential method can be used[94] to determine the band-structures of various semiconductors (tables 25 and 26). A computer program can calculate fully vectorial electronic band structures for materials having diamond or zincblende structures. By using measured energy-gap values for various zincblende crystals, the pseudopotential form-factors of BAs and other compounds have been calculated by means of an iterative technique (table 27).

Table 25. Pseudopotential form-factors (Ryd) of semiconductors with diamond or zincblende structures

Material	V_s^3	V_s^8	V_s^{11}	V_a^3	V_a^4	V_a^{11}
Diamond	−0.690	0.360	0.110	0	0	0
Si	−0.210	0.040	0.080	0	0	0
Ge	−0.230	-0.010	0.060	0	0	0
Sn	−0.200	0.000	0.040	0	0	0
GaP	−0.220	0.030	0.070	0.120	0.070	0.020
GaAs	−0.230	0.010	0.060	0.070	0.050	0.010
GaSb	−0.220	0.000	0.050	0.060	0.050	0.010
InP	−0.230	0.010	0.060	0.070	0.050	0.010
InAs	−0.220	0.000	0.050	0.080	0.050	0.030
InSb	−0.200	0.000	0.040	0.060	0.050	0.010
ZnS	−0.220	0.030	0.070	0.240	0.140	0.040
ZnSe	−0.230	0.100	0.060	0.240	0.140	0.040
ZnTe	−0.220	0.000	0.050	0.130	0.100	0.010
CdTe	−0.200	0.000	0.040	0.150	0.090	0.040
AlSb	−0.210	0.020	0.060	0.060	0.040	0.020
AlPb	−0.204	0.030	0.262	0.053	0.123	0.229
AlAs	−0.221	0.025	0.070	0.080	0.050	−0.004
BP	−0.377	0.093	0.099	0.010	0.020	0.034
BAs	−0.231	0.396	0.094	0.165	0.080	0.010
3C-SiC	−0.385	0.118	0.124	0.160	0.005	0.003
cubic-GaN	−0.226	0.001	0.157	0.305	0.219	0.06

Table 26. Energy gaps (eV) of BAs and BP at Γ, X, L symmetry points

Material	Energy Gap	measured	calculated
BAs	E_g^{Γ}	3.328	3.33
BAs	E_g^{X}	1.444	1.46
BAs	E_g^{L}	2.600	2.59
BP	E_g^{Γ}	5.250	5.22
BP	E_g^{X}	2.180	2.21
BP	E_g^{L}	3.400	3.46

In order to calculate the form-factors it is necessary to begin with average or approximate symmetrical form-factors for the component elements and then vary the asymmetrical and symmetrical form-factors until the simulated energy-gap is close to the measured value (table 28). The form-factors were then used to produce electronic-band diagrams for the compounds. The differences between the measured and calculated energy-gaps at the Γ, X and L symmetry-points were less than 0.05eV.

Calculations showed that pristine BAs monolayers have a graphene-like structure and that the chemical bonding is a mixture of covalent and ionic. The electronic properties are such that a BAs monolayer is a paramagnetic semiconductor with a direct band-gap of 0.682eV. The possibility of vanadium, chromium and manganese doping is governed by positive binding energies of 1.592, 1.668 and 1.876eV, respectively. The incorporation of vanadium and chromium generates half-metallicity in the BAs monolayer, with a spin-down band-gap of 0.873 and 0.897eV, respectively. A monolayer which is doped with manganese is nearly half-metallic. Doping also induces magnetism, and this arises mainly from the 3d orbitals of the transition metals, with a small contribution being to the nearest-neighbour arsenic atoms. In order to tune single-atom catalysis for the purpose of nitrogen reduction, a density functional theory investigation was made[95] of various implanted transition metals in boron-arsenide, boron-phosphide and boron-antimony. The W-BAs system exhibited a high catalytic activity and excellent selectivity: there was a barrier of just 0.05eV along the distal pathway, and a kinetic barrier of only 0.34eV.

The band-gap of about 1.85eV, combined with a large carrier mobility, makes BAs a potential photo-electrode material for solar energy harvesting. Spin-polarized density functional theory calculations were used[96] to investigate the possibility of using surface-functionalized BAs as a photocathode for solar-driven CO_2 reduction. Attention was confined to the (110) face in aqueous solution, and examination of the band-edge positions indicated a high thermodynamic driving force for CO_2 reduction by photo-excited electrons from the conduction band. The catalytic activity of BAs which was functionalized with the co-catalyst, pyridine, was studied with regard to the reaction-path of a hydride-transfer process arising from adsorbed 2-pyridinide. Formation of the latter, and CO_2 reduction, was thermodynamically and kinetically favoured and it was predicted that pyridine-functionalized BAs would be a good candidate for heterogeneous CO_2 photo-reduction.

Density functional theory calculations were used[97] to assess the potential of hexagonal BAs nanosheet as a host material for Li-S batteries. The latter is among the leading candidates for high-performance energy storage, because of its high theoretical capacity (1674mAh/g) and energy-density (2600Wh/kg) as compared with that (300Wh/kg) of Li-ion batteries. High cell polarization and the shuttle effect unfortunately represent a barrier to the commercial use of Li-S batteries. The binding and electronic characteristics of lithium polysulfides, adsorbed on hexagonal BAs surfaces were studied. The results demonstrated the suitability of hexagonal BAs monolayers as a host material, with the binding energies of various lithium polysulfides varying from 0.47 to 3.55eV. The defect surface of the hexagonal BAs monolayer offered optimum binding energies with lithium polysulfides.

In the absence of anharmonic phonon scattering, crystals might exhibit infinite or super intrinsic bulk thermal conductivity. Anharmonic phonon scattering-rates usually increase with temperature and cause a decrease in the thermal conductivity. It was shown[98] that the increase in scattering rates is due to an increase in the phonon population. The anharmonic force constant, or phonon scattering cross-section, decreased with temperature in most solids. The decrease in scattering cross-section compensated for the increase in population, slowed the power-law increase in scattering-rates and decreased the thermal conductivity. The use of a temperature-dependent scattering cross-section increased the thermal conductivity with respect to the 0K ground-state scattering cross-section. A temperature-dependent scattering cross-section could result in a more than 15-fold increase in thermal conductivity at 2000K as compared to that predicted by using the ground-state scattering cross-section. Four-phonon scattering becomes prominent at temperatures greater than 1500K. The universality of the above result was checked by testing silicon, germanium and wurtzite boron arsenide. The same effect was observed. It was very small for strongly harmonic systems. Classical interatomic potentials were used as they describe thermal transport in various materials, and the same conclusion was again drawn. The effect in w-BAs was small. Based on these observations.it was concluded that the temperature softening effect was perhaps stronger for more strongly anharmonic materials, and it increased with temperature.

High-order anharmonic phonon-phonon interactions play an important role in materials like cubic boron arsenide, where a wide phononic energy gap is a critical factor with regard to 4-phonon scattering. By solving the Boltzmann transport equation it was shown[99] that 4-phonon scattering had a marked effect upon thermal transport in honeycomb-structured monolayered BAs and its hydrogenated bilayer form. The lattice thermal conductivity of these structures decreased after taking account of 4-phonon scattering. A drop in lattice thermal conductivity of up to 80% occurred in monolayer BAs, as compared to the case without 4-phonon scattering, and this was due mainly to the suppression of phonon lifetimes. Unlike the case of graphene, the lattice thermal conductivity of monolayer BAs was anomalously lower than that of the hydrogenated bilayer form. This was attributed to the much higher phonon-scattering rate in monolayer BAs as compared to that in hydrogenated bilayer BAs. Comparison of BAs sheets, with and without horizontal mirror symmetry, showed that the contribution of flexural acoustic phonon led to the most marked reduction in both monolayer BAs and hydrogenated bilayer BAs with horizontal mirror symmetry upon including 4-phonon scattering.

Heterostructures

Predictive atomistic calculations were used[100] to investigate the properties of BAs heterostructures, such as the effect of strain upon band-alignment and carrier mobility. The arsenide was envisaged as being both a thin film and a substrate for lattice-matched materials. It was found that isotropic biaxial in-plane strain decreased the band-gap, regardless of its sign or direction. A biaxial tensile strain of 1% increased the in-plane electron and hole mobilities, at 300K, by more than 60% with respect to strain-free values. This was due to a reduction in the electron effective mass and of hole interband scattering. The arsenide could be almost exactly lattice-matched to the opto-electronic tunable-bandgap semiconductors, InGaN and $ZnSnN_2$, by means of alloying and cation-disordering, respectively. The results predicted type-II band-alignments and the absolute band-offsets of the nitrides with respect to BAs. The ultra-high thermal conductivity and intrinsic p-type characteristics of the arsenide, with its high strain-affected electron- and hole-mobilities, as well as its possible lattice-matching and type-II band-alignment with the intrinsically n-type InGaN and $ZnSnN_2$ indicated the promise of BAs heterostructures for electronic and opto-electronic use.

First-principles calculations have been used[101] to investigate the thermal, mechanical, electronic and optical properties of heterostructures which comprised boron arsenide and

WX_2, where X was sulfur or selenium (tables 27 to 30). The binding energy of 289.7meV and 484.6meV, for BAs/WS_2 and BAs/WSe_2, respectively, together with phonon spectra, molecular dynamics and elastic deformation resistance, indicated that the heterostructures were stable; structurally, dynamically and mechanically. These van der Waals heterostructures were all direct-bandgap semiconductors (0.6eV and 0.7eV, respectively). The BAs/WS_2 heterostructure had a type-II band alignment which encouraged the separation of photogenerated carriers and greatly prolonged their lifetime. The BAs/WSe_2 heterostructure exhibited a type-I band alignment which facilitated the rapid recombination of photogenerated carriers. Both heterostructures had a high carrier mobility of 100 to 1000cm2/Vs and an optical absorptivity of about 10^5/cm within the range between ultraviolet and visible. This rendered them very efficient with regard to solar energy. The band structures and carrier mobilities of BAs/WX_2 heterostructures were greatly affected by spin–orbit coupling effect. An external electric field could tailor the band structures, including the transition between direct and indirect band-gaps and changes between type-I and type-II band alignments.

Table 27. Bader charge analysis of BAs, WX_2 monolayers and BAs/WX_2 heterostructures

Material	Component	Charge (electrons)	Charge Transfer (electrons)
BAs	B	2.880	0.120
BAs	As	5.120	
WS_2	W	4.811	1.189
WS_2	S	6.594	
WSe_2	W	5.060	0.940
WSe_2	Se	6.470	
BAs/WS_2	BAs	7.973	0.027
BAs/WS_2	WS_2	10.027	
BAs/WSe_2	BAs	7.984	0.016
BAs/WSe_2	WSe_2	18.016	

Table 28. Elastic constants C_{11}, C_{12} deduced from the strain-energy relationship, and other properties

Parameter	BAs/WS$_2$	BAs/WSe$_2$
C_{11}	277	305
C_{12}	76	56
G (N/m)	101	125
E (N/m)	256	295
γ_{2D} (N/m)*	177	181
υ**	0.275	0.182
σ_{int} (N/m)***	28.4	32.8
D (eV)****	78.94	96.55

*2-dimensional layer modulus, **Poisson ratio,
***intrinsic strength, ****bending modulus

Table 29. Effective mass, elastic modulus, deformation potential constant and carrier mobilities of electrons and holes in BAs/WX$_2$ heterostructures along the transport direction at 300K. Calculated by using the Perdew-Burke-Ernzernhof functional without considering the strong orbit coupling effect

Parameter	Direction	Entity	BAs/WS$_2$	BAs/WSe$_2$
m_0	x	electron	0.337	0.243
m_0	x	hole	-0.304	-0.457
m_0	y	electron	0.266	0.249
m_0	y	hole	-1.063	-1.771
G (N/m)	x	electron	128.49	117.67
G (N/m)	y	electron	125.59	114.89
D (eV)	x	electron	-7.60	-1.48
D (eV)	x	hole	-2.91	-2.16
D (eV)	y	electron	-7.08	-1.46
D (eV)	y	hole	-2.17	-1.68
μ (cm$_2$/Vs)	x	electron	94	3813
μ (cm$_2$/Vs)	x	hole	374	261
μ (cm$_2$/Vs)	y	electron	134	3765
μ (cm$_2$/Vs)	y	hole	189	108

Table 30. Effective mass, elastic modulus, deformation potential constant and carrier mobilities of electrons and holes in BAs/WX$_2$ heterostructures along the transport direction at 300K. Calculated by using the Perdew-Burke-Ernzernhof functional and considering the strong orbit coupling effect

Parameter	Direction	Entity	BAs/WS$_2$	BAs/WSe$_2$
m$_0$	x	electron	0.405	0.237
m$_0$	x	hole	-0.310	-0.577
m$_0$	y	electron	0.490	0.355
m$_0$	y	hole	-1.060	-1.545
G (N/m)	x	electron	124.64	115.55
G (N/m)	y	electron	125.21	115.55
D (eV)	x	electron	-7.28	-1.45
D (eV)	x	hole	-2.83	-1.41
D (eV)	y	electron	-6.86	-1.36
D (eV)	y	hole	-2.06	-2.22
μ (cm$_2$/Vs)	x	electron	55	3408
μ (cm$_2$/Vs)	x	hole	373	455
μ (cm$_2$/Vs)	y	electron	52	2665
μ (cm$_2$/Vs)	y	hole	207	78

Density functional theory and non-equilibrium Green-function based first-principle were used[102] to make an in-depth analysis of infinitely-long boron plus group-V (N, P, As, Sb) linear atomic chains under a tensile stress. This revealed the presence of dative bonds among the atoms of the atomic chains, and a loss of stability of the structures under an applied stress. The boron phosphide linear atomic chain exhibited a superior resistance to tensile stress than did the other chains. The boron arsenide and boron antimonide linear atomic chains exhibited very good electrical and thermal transport, as compared to those of boron nitride and boron phosphide.

Attempts were made[103] to predict the properties of van der Waals heterostructures which consisted of boron arsenide and boron phosphide sheets. Theoretical results showed that BP intercalation into BAs/BP sandwiched heterostructures led to an unusual semiconductor-metal transition which resulted in a dramatic increase in the electronic current; which was very different to that for pure BAs and BP sheets. When a small BP nanotube was inserted into a larger BAs nanotube, relative rotation of the inner and outer tubes also led to a similar semiconductor-metal transition. First-principles calculations were used[104] to investigate the thermal, mechanical, electronic and optical properties of heterostructures that comprised boron arsenide and WX_2, where X was sulfur or selenium. The binding energies were 289.7 and 484.6meV for BAs/WS_2 and BAs/WSe_2, respectively. The phonon spectra, molecular dynamics and elastic deformation resistance data indicated that the heterostructures were stable; structurally, dynamically and mechanically. These van der Waals heterostructures were direct-bandgap (0.6 and 0.7eV, respectively) semiconductors. The BAs/WS_2 van der Waals heterostructure had a type-II band alignment which encouraged the separation of photo-generated carriers and greatly prolonged their lifetime. The BAs/WSe_2 van der Waals heterostructure had a type-I band alignment which facilitated the rapid recombination of photo-generated carriers. Both of the heterostructures exhibited a high (100 to $1000cm^2/Vs$) carrier mobility and high ($10^5/cm$) optical absorptivity between the ultraviolet to visible regions. The band structures and carrier mobilities of the heterostructures were greatly affected by spin–orbit coupling. An external electric field could be used to tailor the band structures, including the occurrence of the transition between the direct and indirect band-gaps and the evolution between type-I and type-II band alignments.

These 2-dimensional materials present new possibilities for 2-dimensional ultra-thin excitonic solar cells. The construction of van der Waals heterostructures is an effective means of combining the properties of single-layer 2-dimensional materials so as to provide an excellent performance. The properties of van der Waals heterostructures of hexagonal BP joined with hexagonal BAs in 2-dimensional excitonic solar cells were studied[105] via a systematic investigation of the electronic and optical properties of heterogeneous structures by means of density functional theory and first-principles calculations. This showed that the heterogeneous structure possessed good opto-electronic properties. These included a suitable direct band-gap and excellent optical absorption. Calculation of the phonon spectrum confirmed the well-defined kinetic stability of the heterostructure. The heterogeneous structure was designed as a model for solar cells, and calculations were made of its solar cell power-conversion efficiency. The latter could attain up to 16.51%; higher than the highest (11.7%). efficiency reported for organic solar cells. This illustrated the potential of such heterostructure as candidates for high-efficiency 2-dimensional excitonic solar cells.

Theoretical calculations indicated that the heterostructure had the lowest binding energy and better kinetic stability. The structures constituted a direct-bandgap semiconductor with a band-gap of 0.93eV and a type-II band alignment which permitted efficient spatial separation between photo-generated electrons and holes. The structures could withstand compressive stresses up to a limit of 4% strain. Under tensile stresses, the structures could be regulated over a wide range. The band-gap of the structures increased with increasing tensile strain. The heterostructure had a very high absorption coefficient of 10^5/cm in the ultraviolet to visible range. The heterostructure also has a small conduction-band offset, a large valence-band offset and the aforementioned high power-conversion efficiency.

Amorphous Boron Arsenide

The short-range order and electrical properties were predicted[106] by means of first-principles molecular dynamics simulations. The amorphous model assumed rapid solidification of the melt and consisted of boron-rich and arsenic-rich domains, with the average coordination numbers of boron and arsenic atoms being 4.97 and 3.34, respectively. The boron atoms had a tendency to form pentagonal pyramidal configurations of the sort that are commonly seen in boron and boron-rich materials. No B_{12} molecules developed in the system, but B_{10} clusters existed in the network. The arsenic atoms tended to form chains and 4-membered rings. Amorphization led to an approximately 31% volumetric expansion of the network. The results indicated the existence of a marked chemical disorder, and a short-range order which was considerably different to that of the crystal.

Cubic Boron Arsenide

The pseudopotential plane-wave approach, density functional theory, density functional perturbation theory and the generalized gradient approach for the exchange-correlation functional were used[107] to calculate the structural phase stability, elastic constants and thermodynamic properties of boron arsenide. The latter transformed from a zincblende phase to a rock-salt phase (table 31) at a transition pressure of 141.2GPa, with a volume contraction of about 8.2%. The thermodynamic properties at pressures of up to 125GPa and temperatures of up to 1200K were determined. The calculated melting temperature of 2116K differed from the theoretical value of 2132.83K by only 0.8%, and the difference between the calculated value, 1.86, of the Grüneisen parameter and the theoretical value of 1.921 was only about 3.2%.

The thermal properties of bulk zincblende BAs were investigated[108] by means of *ab initio* calculations which were performed by using the local density approximation and generalized gradient approximation methods for the exchange-correlation potential. Phonon dispersion relationships were studied within the framework of density functional perturbation theory, and the thermal properties were computed by using the quasi-harmonic approximation. The temperature dependence of the lattice constant, isothermal bulk modulus, linear thermal expansion coefficient, average Grüneisen parameter and specific heat capacity at constant pressure were calculated (table 32). The calculated values at 300K were in excellent agreement with the reported experimental values of 148GPa, 3.85 x 10^{-6}/K, 0.82 and 0.4J/gK, respectively.

Table 31. Calculated properties of boron arsenide

Property	Value	Units
Isothermal bulk modulus	142.8	GPa
Linear thermal expansion coefficient	4.01 x 10^{-1}	/K
Average Grüneisen parameter	0.816	
Specific heat capacity at constant pressure	0.34	J/gK

Table 32. Miscellaneous properties of boron arsenide

Property	Value
Structure	zincblende, cubic, F$\bar{4}$3m
Lattice constant	4.78A
Band-gap	1.82eV
Refractive index (657nm)	3.29
Refractive index (908nm)	3.04
Density	5.22g/cm^3
Sound velocity (longitudinal) <100>	7390m/s
Sound velocity (longitudinal) <111>	8150m/s
Sound velocity (transverse) <100>	5340m/s
Thermal conductivity	1300W/mK
Volumetric heat capacity	2.09J/cm^{3}K
Thermal expansion coefficient (linear)	3.85 x 10^{-6}/K
Thermal expansion coefficient (volume)	11.55 x 10^{-6}/K

In order to estimate the physical characteristics of cubic boron arsenide, Keating's model of force constants and Harrison's method of bonding orbitals have been used[109]. A comparison of the results with experimental data and with other calculations confirmed the applicability of the approach.

In order to exploit fully the high conductivity of BAs there must be a compatible coefficient of thermal expansion between a heat-sink and the semiconductor in order to minimize thermal stresses. An experimental study was therefore made[110] of the coefficient of thermal expansion of BAs at temperatures ranging from 100 to 1150K (table 33) by using a combination of X-ray single-crystal diffraction and neutron powder diffraction methods. The room-temperature coefficient of thermal expansion was 3.6 x 10^{-6}/K.

*Table 33. Lattice parameter of BAs between 100 and 1500K
determined using X-ray or neutron diffraction methods*

Temperature (K)	a (Å)
100	4.7741
120	4.7743
140	4.7743
160	4.7746
180	4.7749
200	4.7751
220	4.7753
240	4.7756
260	4.7759
280	4.7762
299.8	4.7766
300	4.7766
320	4.7770
340	4.7774
360	4.7780
380	4.7784

400	4.7789
401.2	4.7784
420	4.7792
440	4.7797
460	4.7801
467.2	4.7799
543.7	4.7821
603.3	4.7832
667.6	4.7859
743.3	4.7873
779.9	4.7891
848.4	4.7909
919.2	4.7933
1042.8	4.7971
1097.5	4.7991
1168.4	4.8015

Some theoretical calculations of the thermal expansion coefficient and Grüneisen parameter had yielded inconsistent results, and so the linear thermal expansion coefficient was deduced[111] from new X-ray diffraction measurements which were made between 300 and 773K. These results were in good agreement with first-principles calculations that took account of atomic interactions up to the fifth-nearest neighbours. Using the measured thermal expansion coefficient and specific heat, the Grüneisen parameter was found to be 0.84 at 300K. This was in very close agreement with the value of 0.82 which had been calculated using first-principles methods and was much lower than previous theoretical predictions. The results again confirmed that BAs permitted a better thermal-expansion compatibility with other semiconductors than did high-conductivity materials such as diamond and cubic boron nitride. The thermal expansion coefficient was higher than that of previous first-principles predictions which had assumed short-range anharmonic force-constants. It is found that extension up to fifth-nearest neighbour was indeed necessary to produce first-principles calculations that agreed with measurements. The calculated room-temperature thermal expansion parameter (4.0×10^{-6}/K) was then close to the measured value of 4.2×10^{-6}/K. The differences between calculated and measured results were within experimental error between 300 and 773K. It was concluded that long-range atomic interaction played an important role in the lattice anharmonicity of BAs.

The finite-temperature opto-electronic properties were studied[112] by considering both electron-phonon coupling and thermal expansion. The inclusion of electron-phonon coupling was essential in order to capture the temperature-dependence of the opto-electronic properties, whereas thermal expansion made a negligible contribution to the highly covalent bonding. It was predicted that, with increasing temperature, the onset of optical absorption was subject to a red-shift and the absorption peaks became smoother. The phonon-assisted absorption at energies below those of the optical gap had a coefficient that was between 10^{-3} and 10^{-4}/cm. A good agreement with the measured indirect band-gap was obtained only if exact exchange, electron-phonon coupling and spin-orbit couplings were all included in the calculations. An investigation was made[113] of the direct and indirect optical absorption of BAs and BSb by using first-principles calculations. The onset of absorption which corresponded to the size of the indirect band-gap was deduced by considering the phonon-assisted second-order indirect optical absorption. The temperature-dependent calculations also indicated the red-shift of the onset of absorption, and the enhancement and smoothness of the optical absorption spectra. In order to introduce first-order absorption in the visible range, the doping effect of congeners was considered without assuming the assistance of phonons. This showed that a decrease in the local direct band-gap arose either from a decrease in the bonding-antibonding repulsion of p orbital states by heavier group-III elements or from a similar effect that was exerted by lighter group-V elements on the s orbital states.

A time-resolved ultra-fast quasi-particle dynamics investigation was made[114] of excited-state ultra-fast relaxation channels dictated by electron-phonon coupling, phonon-phonon scattering and radiative electron-hole recombination. These were unambiguously identified, together with their typical interaction times. During the analysis the fluence-dependence method, for obtaining the electron-phonon coupling strength, was generalized to room temperature. The electron-phonon coupling strength was deduced from the dynamics, yielding a value of $\lambda_{T2} = 0.008$; corresponding to $\lambda\langle\Omega^2\rangle = 1.18$/ps. This indicated the occurrence of an unusually weak coupling between the electrons and phonons. It was suggested that the existence of an ultra-small electron-phonon coupling strength could be a requirement for the existence of ultra-high thermal conductivity.

Table 34. Zero-temperature formation energies, E_f, and Arrhenius-fitted formation enthalpies, H_f, at the mid-gap for native point defects in cubic BAs.

Material	Defect	E_f (eV)	H_f (eV)
B-rich	V_{As}^0	4.79	4.773
B-rich	As_B^0	3.36	3.628
B-rich	B_{As}^{1-}	0	0
B-rich	As_B^{2+}	5.36	5.704
B-rich	B_i^+	4.20	4.312
B-rich	As_i^0	8.45	8.635
B-rich	V_{As}^0	7.47	7.624
B-rich	V_B^0	0.68	0.7763
As-rich	B_{As}^{1-}	5.35	5.702
As-rich	As_B^{2+}	0	0
As-rich	B_i^+	6.88	7.163
As-rich	As_i^0	5.77	5.784
As-rich	V_{As}^0	5.98	5.982
As-rich	V_B^0	2.37	2.418
stoichiometric	B_{As}^{1-}	2.37	2.418
stoichiometric	As_B^{2+}	3.04	3.464
stoichiometric	B_i^+	5.29	5.521
stoichiometric	As_i^0	7.50	7.442

Defects

A first-principles study was made[115] of all of the possible native point defects in BAs, showing that antisites are the constitutional defects in off-stoichiometric samples. On the other hand, B_{As} antisites and boron vacancies predominate in stoichiometric samples. It was demonstrated that, under conditions of thermodynamic equilibrium and material stoichiometry, the chemical potentials are functions of the temperature, Fermi level and stoichiometry. This permitted the calculation of the formation energies and equilibrium concentrations of the native point defects (table 34). It was noted that, whereas B_{As}

antisites and boron vacancies maintain stoichiometry under most conditions, As_B replaced V_B in extremely p-type material at low temperatures. The full free energies of formation were calculated for various temperatures and this revealed that differences of up to 14% existed between Arrhenius-fitted formation energies and the 0K total-energy values. This stoichiometry-balancing approach permitted the correlation of chemical potentials with the stoichiometry in a meaningful manner for a real material and offered some quantitative insight into the size of point-defect populations.

Table 35. Shallow donor and acceptor ionization energies in BAs

Defect	Nature	Ionization Energy (eV)
Se_{As}	donor	0.16
Be_B	acceptor	<0.03
Te_{As}	donor	0.13
Si_{As}	acceptor	<0.03
Si_B	donor	0.14
Ge_{As}	acceptor	0.03
Ge_B	donor	0.17
Mg_B	acceptor	0.19

Hybrid density functional theory calculations were used[116] to determine the formation energies and thermodynamic charge transition levels of native point defects, impurities and shallow dopants. It was found that As_B antisites, boron-related defects (V_B, B_{As}), B_i-V_B complexes, and antisite pairs were the predominant intrinsic defects. Native B_{As} was expected to exhibit p-type conduction due to the acceptor-type characteristics of V_B and B_{As}. Carbon substitutional defects and hydrogen interstitials had relatively small formation energies and were expected to contribute free holes. Interstitial hydrogen was also found to be stable in the neutral charge state, while Be_B, Si_{As} and Ge_{As} were predicted to be very good shallow acceptors with low ionization energies (<0.03eV) and negligible compensation by other point defects. Donors such as Se_{As}, Te_{As}, Si_B and Ge_B had a relatively large ionization energy (*circa* 0.15eV, table 35) and were likely to be passivated by native defects such as B_{As} and V_B, as well as C_{As}, H_i and H_B. A hole and electron doping asymmetry arose from the heavy effective mass of the conduction band, due to its boron orbital nature, as well as from boron-related intrinsic defects which compensated donors. The large electron effective mass of BAs, originating from the small radius of the B orbitals, led to relatively large ionization energies for shallow donors.

Mechanical properties

Many fundamental properties of BAs remain relatively unexplored experimentally because high-quality single crystals have only recently become available. The mechanical properties of boron arsenide that was doped with nitrogen were studied[117] by using density functional theory. It was found that $BAs_{25}N_{75}$ was a potential superhard material, given its Vickers microhardness of 40GPa.

The properties of single crystals were studied[118] by combining experimental measurements with first-principles calculations. The Vickers hardness of 22GPa suggested that BAs was a hard, but not superhard, material. The bulk and Young's moduli were measured to be 142 and 388GPa, respectively.

A pico-ultrasonic method, based upon ultra-fast optical-pump probe spectroscopy, was used[119] to measure an elastic modulus value of 326GPa (table 36); twice that of silicon. Temperature-dependent X-ray diffraction was used to measure a linear thermal expansion coefficient of 3.85×10^{-6}K; very close to that of GaN.

Table 36. Experimental physical properties of boron arsenide

Property	Experimental Value	Units
C_{11}	285	GPa
C_{12}	79.5	GPa
C_{44}	149	GPa
S_{11}	3.99	10^{-12}/Pa
S_{12}	-0.87	10^{-12}/Pa
S_{44}	6.71	10^{-12}/Pa
G	325	GPa
S	128	GPa
B	148	GPa
Poisson ratio	0.22	
Gruneisen parameter	0.82	

Materials Research Forum LLC

https://doi.org/10.21741/9781644902233

The room-temperature elastic constants of boron arsenide single crystals were deduced[120] from Brillouin frequencies that were measured using picosecond interferometry. The crystals were cut and polished in various orientations in order to access waves which were travelling in various directions. The Brillouin frequencies of quasi-longitudinal waves were determined in 5 different orientations. Quasi-shear waves were observed in one orientation. The propagation directions and acoustic velocities were used to construct Christoffel equations which were then solved for the elastic constants (table 37). The measured elastic constants differed by less than 5% and 17% from calculated elastic constants that were predicted by using local density approximation and Perdew-Burke-Ernzerhof density functional calculations, respectively.

Table 37. Stiffness parameters of boron arsenide

Parameter	Value (GPa)
C_{11}	285
C_{12}	79.5
C_{44}	149

A first-principles study was made[121] of the temperature-dependence of the elastic constants, and the consistency of the predictions was tested by calculating the temperature-dependent sound velocity of the longitudinal acoustic mode along the [111] direction. This also furnished the room-temperature phonon dispersion, the temperature-dependent thermal expansion, isobaric heat capacity and average Grüneisen parameter. The values of the adiabatic elastic constants at 300K were C_{11} = 2807kbar, C_{12} = 754kbar and C_{44} = 1507kbar. Between 0 and 1800K the softening of the adiabatic elastic constants amounted to about 11% for C_{11}, 9% for C_{12} and 13% for C_{44}. This degree of softening was consistent with the temperature variation of the longitudinal sound velocity, as measured along the [111] direction. The slopes of the curves differed below room temperature but were similar at higher temperatures. The softening was slightly less than that found for silicon at 0 to 800K.

The effect of atomic relaxation on the temperature-dependent elastic constants is usually accounted for at zero temperature by minimization of the total energy at each straining stage. An investigation was made[122] of the order-of-magnitude of this approximation for the C_{44} elastic constant of a zincblende material such as BAs (table 38). The effect of finite-temperature atomic relaxation within the quasi-harmonic approximation was estimated by means of the first-principles computation of the internal strain tensor from the second derivatives of the Helmholtz free-energy with respect to strain and atomic displacement. The case of BAs revealed a clear difference between softening of the temperature-dependent elastic constants as computed using the zero-temperature and finite-temperature atomic relaxations. The softening of C_{44} for BAs passed from 8% to 7% between 0 and 1200K.

Table 38. Correction to C_{44} as calculated using various methods

Parameter	Si	C	SiC	GaAs	BAs
a_0 (au)	10.2065	6.6821	8.1835	10.6085	8.9660
ω_{TO} (/cm)	513	1320	798	269	704
C_{44}^{FI} (kbar)	1060	5971	2840	799	1772
ΔC_{44}^{rel} (kbar)	299	72	313	213	212
ΔC_{44}^{IS} (kbar)	299	74	320	209	213
ξ	0.54	0.13	0.42	0.55	0.38
Λ	0.389	0.173	0.347	0.290	0.347

Rel: atomic relaxation value, FI: frozen-in value, IS: internal strain formalism, ω: phonon angular frequency, Λ: internal strain, ξ: Kleinman parameter

Other uses

Photo-electrodes which are made from boron arsenide can be easily prepared directly from the elements and can convert sunlight into electricity[123]. A p-type boron arsenide photo-electrode was prepared[124] from a thin layer of boron arsenide located on a boron substrate. The structure of the material was determined by using X-ray diffraction and scanning electron microscopy. The surface composition was determined by means of X-ray photo-electron spectroscopy. The electrode was photo-active in visible light and UV-vis radiation and generated a photocurrent of about $0.1mA/cm^2$ under UV-vis irradiation at an applied potential of $-0.25V_{Ag/AgCl}$. Mott-Schottky plots for this electrode revealed the existence of an estimated flat-band potential near to the onset-photopotential. The estimated indirect band-gap, as deduced from incident photon-to-electron conversion efficiency-plots was 1.46eV.

The nitrogen electroreduction reaction which occurs in aqueous solutions under ambient conditions is a possible means for producing ammonia, but it is difficult to develop stable and low-cost catalysts of high efficiency and high selectivity. A density functional theory investigation[125] was made of the possible application of boron-containing cubic boron phosphide or boron arsenide as metal-free nitrogen electroreduction reaction electrocatalysts. It was shown that gaseous N_2 could be activated on the boron-terminated (111) polar surfaces of BP and BAs and be effectively reduced to NH_3 via an enzymatic pathway: with a limiting potential only -0.12V on BP and only -0.31V on BAs. These proposed boron-terminated (111) surfaces possessed a large active region for N_2 reduction and could also significantly inhibit a competing hydrogen-evolution reaction. They therefore exhibited a high efficiency and selectivity with regard to the nitrogen electroreduction reaction.

The efficient control and manipulation of spin degrees of freedom, without the use of a magnetic field, is one of the main aims in the development of spintronic devices. A class of bipolar magnetic semiconductors can be prepared[126] from semi-hydrogenated BAs nanosheets with a Curie temperature of 307K. Both electron self-doping and hole self-doping of the structure with semihydrogenated arsenic atoms can induce a transition from a ferromagnetic semiconductor to a half-metal. Such bipolar magnetic semiconductors can resist the effect of strain or of a strong electric field. This offers a promising method for the electrical manipulation of carrier spin-orientations in 2-dimensional materials.

Thermal conductivity

An early first-principles approach predicted that zincblende boron arsenide would have an ultra-high lattice thermal conductivity of over 2000W/mK at room temperature, thus making it comparable to that of diamond. A further detailed first-principles study[127,128] of the phonon thermal transport in boron arsenide compared its unusual behaviour with that of related materials such as zincblende boron nitride, boron phosphide, boron antimonide and gallium nitride. In spite of the above prediction, subsequent experimental measurements[129] yielded conductivities of only about 200W/mK when single crystals of BAs with a zincblende cubic structure and lattice parameter of a = 4.7830Å were tested. In order to explain the discrepancy, measurements were made[130] of the phonon dispersion of single-crystal BAs along high-symmetry directions, using inelastic X-ray scattering and were compared with first-principles calculations. On the basis of the measured phonon dispersion, it was possible to confirm the theoretical prediction of a large frequency-gap between the acoustic and optical modes, plus a bunching of the acoustic branches. These were considered to be the main reasons for the expected ultra-high thermal conductivity.

Isotopic scattering can play an important role in determining the thermal conductivity of many materials (table 39). Boron arsenide contains a large mixture of boron isotopes, but relatively small contributions arise from isotopic scattering; the contributions are even those available in cubic BN, BSb and diamond. The cause is that, in BAs, the large arsenic-to-boron mass-ratio causes the atomic motion of the short-wavelength acoustic phonons, which make the largest contributions, to be dominated by the greater arsenic mass; which is isotopically pure. Both the intrinsic conductivity and the isotopically controlled conductivity are consequently similar to that of diamond.

It is informative to compare the conductivities of intrinsic and isotopic BAs to those of cubic BN and BSb. The former is a conventional high-conductivity material, with its light constituent atoms and stiff covalent bonding giving it a high room-temperature intrinsic conductivity of some 2100W/mK. On the other hand the intrinsic conductivity of BSb, of about 1200W/mK arises from a lack of scattering.

Table 39. Properties of various materials

Material	$M_{average}$ (amu)	θ_{Debye} (K)	$\kappa_{intrinsic}$ (W/mK)	$\kappa_{isotopic}$ (W/mK)
Cubic-BN[a]	12.41	2025	2145	940
BP[a]	20.89	1110	665	580
BAs[a]	42.87	700	3170	2240
BSb[a,c]	66.28	495	1180	465
Diamond[b]	12.01	2280	3450	2290
Si[d]	28.09	710	155	145
Ge[e]	72.59	415	75	60

Isotope concentrations: a (19.9% ^{10}B, 80.1% ^{11}B), b (98.1% ^{12}C, 1.1% ^{13}C), c (57.2% ^{121}Sb, 42.8% ^{123}Sb), d (92.2% ^{28}Si, 4.7% ^{29}Si, 3.1% ^{30}Si), e (20.4% ^{70}Ge, 27.3% ^{72}Ge, 7.8% ^{73}Ge, 36.7% ^{74}Ge, 7.8% ^{76}Ge)

A Raman spectroscopic and first-principles simulation study[131] of the Raman T_2 phonon involving boron vibrations in ^{nat}BAs revealed a complex isotopic 2-mode behaviour which was unique among isotopically disordered materials. The majority isotope ^{11}B phonon at 704/cm obeyed an average isotopic mass dependence and mass disorder renormalization, while a band at 723/cm arose from predominant ^{10}B vibrations in ^{nat}BAs. An absence of T_2 LO-TO splitting was observed in both ^{nat}BAs and ^{11}BAs, and third-order and fourth-order anharmonicity contributions compensated one another in the temperature-dependence of the T_2 phonon. The unusual vibrational properties of BAs contributed to its weak phonon-phonon scattering and phonon-isotope scattering, which were in turn responsible for its very high conductivity. Cubic boron nitride crystals with controlled amounts of ^{10}B or ^{11}B isotopes have also been measured[132] to have thermal conductivities greater than 1600W/mK at room temperature. It was however found that the isotopic enhancement of the conductivity was considerably smaller for boron phosphide and boron arsenide, as the given isotopic mass disorder became increasingly invisible to phonons.

A Raman scattering study was made[133] of isotopically tailored cubic monocrystals with 11 isotopic compositions ranging from nearly-pure cubic ^{10}BAs to nearly-pure cubic ^{11}BAs. The results showed the effects of a marked mass disorder upon the optical phonons, and the appearance of 2-mode behaviour in the Raman spectra of mixed crystals. The marked isotopic disorder also led to the relaxation of 1-phonon Raman selection rules and resulted in disorder-activated Raman scattering by acoustic phonons.

Four-probe thermal and thermo-electric transport measurements were performed[134] on an individual BAs microstructure that had been synthesized using vapour transport methods. The thermal conductivity decreased slightly with temperature between 250 and 350K, and this temperature-dependence suggested that extrinsic phonon scattering processes played an important role, in addition to intrinsic phonon-phonon scattering. The room-temperature conductivity (186W/mK) was higher than that of bulk silicon but was a factor of 4 lower than the predicted result for defect-free non-degenerate BAs rod having a similar diameter (1.15µm). The measured p-type Seebeck coefficient and thermo-electric power-factor were comparable to those of bismuth telluride. It was suggested that it was necessary to reduce defect and boundary scattering and, in addition, control the electron-scattering of phonons in order to attain the predicted intrinsic lattice thermal conductivity.

Single crystals which contained no detectable defects exhibited a measured room-temperature thermal conductivity of 1300W/mK[135]. A spectroscopic study, combined with atomistic theory, showed that the distinctive band-structure of BAs permits a very long phonon mean free path, plus a marked high-order anharmonicity via the 4-phonon process.

Materials Research Forum LLC

https://doi.org/10.21741/9781644902233

Although crystals had been grown by means of chemical vapor transport, the growth process relied upon spontaneous nucleation and resulted in small crystals comprising multiple grains and containing various defects. A more controllable chemical vapor transport method produced[136] large (400 to 600μm) single crystals by using tiny, but selected, BAs single crystals as seeds. This led to single crystals having a thermal conductivity of 351W/mK at room temperature: almost twice that of previous crystals.

The conductivity of BAs included contributions that arose from phonons having anomalously long mean free paths of the order of 2μm; up to 3 times longer than those of diamond or BN. The conductivity of BAs was therefore sensitive to phonon scattering arising from crystal boundaries. The sensitivity to phonon-scattering by crystal boundaries required the preparation of large (at least 10μm) single crystals in order to avoid large impairments of the conductivity. High-quality samples containing few defects were also to be preferred. The order-of-magnitude lower room-temperature conductivity of zincblende GaN was attributed to the more widely separated acoustic phonon branches, the larger anharmonic force constants and a large isotope mixture which constituted the heavy, rather than the light, atomic component. That is, the high conductivity of BAs was due to bunching of the acoustic phonon branches, to the large mass-ratio of the constituent atoms and to an isotopically pure heavy atom. None of these features were expected to be associated with a high conductivity. The sensitivity of conductivity to changes in the vibrational properties suggested that the conductivity might be further improved by fine-tuning these features. Manipulation, perhaps by straining, of the chemical bonding of a large mass-ratio compound might bunch together the acoustic branches and force the conductivity up to yet-higher values.

The cubic material exhibits high intrinsic electron/hole mobilities, and ionized impurity scattering plays a more important role in carrier scattering than do other scattering processes. These mobilities could be increased[137] by 14.9 and 76.2% for electrons and holes, respectively, by the use of strain-engineering. An investigation of the opto-electronic properties of this indirect semiconductor, by considering the many-body excitonic effects, revealed that the contribution arising from finite-momentum excitons with regard to optical properties was greater for photon-energies ranging from 2.25 to 3.50eV, as compared to that arising from zero-momentum excitons. The phonon-electron coupling to the total lattice thermal conductivity was non-trivial at low temperatures.

Cubic boron arsenide was believed to incorporate high concentrations of crystal imperfections that could impair the thermal conductivity and act as sources of unintentional p-type conductivity. This behavior was generally attributed to the presence of native defects, but optical and magnetic resonance spectroscopy together with first-principles calculations suggested[138] that unintentional acceptor impurities such as silicon and/or carbon were more likely culprits in causing the observed conductivity. This suggestion also showed that the true low-temperature band-gap of the cubic arsenide was 0.3eV higher than the generally accepted value of about 1.5eV. Low-temperature photoluminescence measurements revealed impurity-related processes such as donor-acceptor pair recombination, while electron paramagnetic resonance revealed evidence of effective mass-like shallow acceptors. Hybrid density functional calculations indicated that native defects were unable to produce such signals. It was instead found that group-IV impurities were easily incorporated at arsenic sites and acted as shallow acceptors. Such impurities could dominate the electrical properties and their influence on phonon-scattering was to be considered with regard to the thermal conductivity.

By using a first-principles approach it was shown[139] that vacancies lead to an anomalously large suppression of the lattice thermal conductivity of the cubic arsenide. This effect was linked to the unusually long phonon lifetimes in BAs and resulted in a greater reduction in the conductivity than that which occurs in diamond. The large changes in bonding which occur around vacancies could not be adequately treated by using standard perturbation methods and were instead handled by using an *ab initio* Green function approach. The arsenic and boron vacancies were found to exert similar effects upon the conductivity. It was also shown that the commonly used mass-disorder models which were applied to vacancies failed to work in the case of large mass-ratio compounds materials such as BAs. This was because they incorrectly predicted a much stronger phonon-scattering when the vacancy was on the heavy-atom site and a much weaker phonon-scattering when the vacancy was on the light-atom site.

Although the electron/hole mobilities had been calculated, there was an absence of experimental data on the electronic properties. A photoluminescence study was therefore made[140] of single crystals at various temperatures and pressures. This revealed an indirect band-gap and 2 donor–acceptor pair recombination transitions. Using first-principles calculations and time-of-flight secondary-ion mass spectrometry the 2 donor–acceptor pair transitions were shown to originate from silicon and carbon impurities which occupied shallow energy levels in the band-gap. High-pressure photoluminescence spectra showed that the donor level with respect to the conduction band minimum decreased with increasing pressure. This then affected the release of free carriers from defect states. These results again confirmed the possibility of using strain-engineering to manipulate the transport properties of BAs.

The existence of defects markedly limits the thermal conductivity of BAs and also changes its pressure-dependent thermal transport behavior. Picosecond transient thermoreflectance and femtosecond time-domain thermoreflectance techniques revealed[141] a non-monotonic dependence of the thermal conductivity upon pressure. This was not caused by pressure-modulated phonon-phonon scattering, because that was predicted to cause only 10 to 20% change in the conductivity. It was instead attributed to various competing effects such as defect-phonon scattering and the modification of structural defects by high pressures. The thermal conductivity can be deduced from

$$\kappa = 1/3Cv^2\tau$$

where C is the heat capacity, v is the group velocity and τ is the phonon lifetime. The overall thermal conductivity is the summation of all of the phonon modes. Compression which is applied by a diamond anvil cell can reduce the interatomic distance of the BAs lattice and increase its interatomic bonding strength. The heat capacity depends moreover on the density. When the interatomic distance is decreased by pressure, the density and heat capacity increase and stronger interatomic bonds produce a higher phonon group velocity. Meanwhile, stiffer bonding increases the anharmonic scattering among phonons; decreasing the phonon lifetime and therefore the thermal conductivity. When the pressure reaches a certain point, and the interatomic distance is small, it becomes more and more difficult to compress the material further; due to the repulsive component of the Coulomb interaction between electrons. The combined effect of these changes could result in an initial increase, followed by saturation, as the pressure increased. This is habitually observed in covalent solids and in 2-dimensional van der Waals materials, unless a phase transition intervenes. The initial increase which is observed in BAs is consistent with the behaviour of other materials. There is however a decrease in the case of BAs as the pressure continues to increase. One possible explanation for this non-monotonic behaviour is the occurrence of a phase transition. When the interatomic distance exceeds some threshold value, the energy which is accumulated in some materials can become very high. If another phase of lower energy exists, there will be a rearrangement of the atoms so as to form the alternative phase. A sign of the emergence of a new phase is the appearance of new phonon-modes. A previous study of the Raman spectra of BAs at up to 20GPa had revealed a fine splitting of the LO and TO phonons of ^{11}B. No new phonon modes were observed in BAs under high pressures, suggesting that no phase transition occurs in BAs within that pressure range.

Other factors to be considered are pressure-modulated phonon-phonon interactions and phonon–defect interactions. It had been predicted that there was an increase and then decrease in the thermal conductivity with pressure that was caused by the competition of 3-phonon and 4-phonon scattering processes. The 4-phonon scattering was important in BAs due to the large energy-gap between the acoustic and optical phonon branches.

When the 3-phonon processes could no longer satisfy energy-conservation, 4-phonon processes were required. The 4-phonon scattering-rates decrease with pressure and thus increase the thermal conductivity whereas 3-phonon scattering-rates increase with pressure and tend to decrease the conductivity. In the pressure range below 20GPa, the 4-phonon reduction effect predominated and led to an overall increase in thermal conductivity with pressure. At pressures greater than 20GPa, the 3-phonon scattering increase predominated and thus the overall conductivity decreased. A peak thermal conductivity occurred at about 17.5GPa. The pressure-dependent behavior of the thermal conductivity of as-grown BAs samples can thus exhibit complex features which result from a competition between various physical mechanisms. It can be proposed that there are 5 possible mechanisms which can effect this behavior:

Competition between 3-phonon and 4-phonon scattering. Competition between various channels of 3-phonon scattering. Competition between defect–phonon and phonon–phonon scattering. Temporary pressure-modification of structural defects, which is reversed when the pressure is removed. The first two mechanisms consider only the intrinsic phonon–phonon scattering in perfect BAs crystals which can cause conductivity changes of less than 20%. The observed behavior mainly reflected the effect of point defects and structural defects.

First-principles calculations were used[142] to identify a phenomenon which had never been predicted or previously observed. This was the effect of the competing responses of 3-phonon and 4-phonon interactions with regard to pressure increases; thus leading to a non-monotonic pressure-dependence of the thermal conductivity. The latter first increased in the expected manner, and then decreased. The resultant peak in the conductivity exhibited a strong temperature-dependence due to a rapid strengthening with temperature of the 4-phonon interactions relative to 3-phonon processes. This permitted pressure to be used as a means for tuning the interplay between the competing phonon-scattering mechanisms.

The compression of boron arsenide leads to marked changes in the phonon dispersion. This permits the evaluation of first-principles theories of how phonon dispersion affects the 3- and 4-phonon scattering rates. The thermal conductivity was measured[143] from 0 to 30GPa by using time-domain thermoreflectance and a diamond anvil cell. The material was observed to have a pressure-independent thermal conductivity below 30GPa, and it was concluded that this revealed an important relationship to exist between phonon-dispersion properties and 3- and 4-phonon scattering rates. The thermal conductivity as a function of pressure was determined for two samples having ambient thermal conductivities of 350 and 480W/mK. Monocrystalline MgO was used as a control. The thermal conductivities of both samples depended only slightly upon pressure between 0 and 25GPa. Such a pressure-independent thermal conductivity between 0 and 25GPa is atypical for non-metallic materials; simple models for phonon thermal conductivity predict a monotonic increase in conductivity with increasing pressure. Three-phonon-scattering rates are governed by phonon anharmonicity, and the latter usually decreases with increasing pressure. In most non-metallic materials, 3-phonon scattering rates are expected to decrease with increasing pressure. A relaxation-time approximation model was used to analyze the experimental results and determine how the pressure produced changes in the phonon group velocities, phonon-phonon scattering and defect-scattering affected the conductivity. This analysis suggested that the weak pressure-dependence of the conductivity existed because the total phonon-phonon scattering rates were pressure-independent. The results were consistent with density functional theory predictions that acoustic bunching, 3-phonon scattering and 4-phonon scattering govern the conductivity. In the case of both samples the conductivity deviated from the overall trend at pressures of between 4 and 6GPa, and this was tentatively attributed to appreciable changes in the optical properties at 1.58eV which occurred across this pressure range. The optical properties near to 1.5eV were greatly affected by absorption, due to interband transitions. The interband transition energy threshold increased under pressure, thus causing the thermoreflectance at 1.58eV to pass through zero and change sign. The thermoreflectance also flipped sign at about 6GPa.

The presence of defects greatly limits the thermal conductivity of BAs and also changes the pressure-dependent thermal transport behavior . By using picosecond transient thermoreflectance and femtosecond time-domain thermoreflectance method, a non-monotonic dependence of the thermal conductivity on pressure was revealed[144]. This was not caused by pressure-modulated phonon-phonon scattering, which was expected to change the thermal conductivity by only 10% to 20%, but was attributed to several competing effects such as defect-phonon scattering and modified structural defects under high pressures. A more detailed list identified 5 possible mechanisms. These were competition between 3-phonon and 4-phonon scattering, competition between various channels of 3-phonon scattering, competition between defect–phonon and phonon–phonon scattering and temporary modification of structural defects.

Phonon anharmonicity is critical to the accurate prediction of a material's thermal conductivity. Calculations which are based upon perturbation theory are however difficult and time-consuming; especially for high-order phonon scattering. Cubic boron arsenide was used as an example of combining the machine learning potential with molecular dynamics simulations in order to predict the conductivity and evaluate the effect of anharmonicity upon thermal transport. A machine learning potential, based upon the matrix tensor algorithm, was developed[145] which could accurately describe the lattice dynamics. Analysis of the phonon spectral energy density revealed that a machine learning potential could capture both the phonon mode softening and linewidth broadening which were produced by the anharmonicity at finite temperatures. The BAs exhibited a strong anharmonicity, as revealed by a large deviation from the equilibrium position and a pronounced phonon broadening; especially at high temperatures. Based upon the phonon Boltzmann transport equation and a 3-phonon scattering process, calculations demonstrated that the accuracy of the machine learning potential in predicting conductivity is comparable to that offered by density-functional theory calculations for BAs. This framework greatly overestimated the conductivity of BAs as compared with experimental results. This was attributed to the marked impact of high-order phonon scattering in BAs. On the other hand, the conductivities which were predicted by equilibrium molecular dynamics simulations, combined with machine learning potential, agreed well with experimental values.

An investigation was made[146] of the effects of impurities upon the thermal and electrical properties of cubic BAs. Time-of-flight secondary ion mass spectrometry and electron probe microanalysis revealed the presence of impurities such as silicon, carbon, oxygen, hydrogen, tellurium, sodium and iodine. The silicon, carbon and hydrogen could serve as shallow acceptors, thus leading to p-type conduction usually being found for BAs. The thermal conductivity and hole mobility were reduced in samples with higher impurity concentrations, due to the enhanced impurity-scattering of phonons and holes, respectively. First-principles calculations predicted the thermal conductivity reductions which would be produced by various impurities. The substitution of oxygen for arsenic led to a large bond-distortion, resulting from the breaking of T_d symmetry. This led to unusually marked phonon-scattering with a correspondingly large reduction in the thermal conductivity. It was assumed that there was only one type of dopant, that each dopant contributed one hole and that all of the dopants were fully ionized at 300K. The estimated impurity concentrations were 0.0150at% for sample s1 and 0.145at% for sample s2 (table 40, figures 8 to 10). The measured thermal conductivity decreased with increasing impurity concentration. Over most of the impurity-concentration range, the conductivity changed slowly with concentration and doubling the density of impurities did not double the thermal resistivity. It was deduced that the same amount of impurity might affect the thermal conductivity less if the impurities were concentrated within a part of the material rather than being homogeneously distributed. It was proposed that high-purity boron and arsenic should be used as starting materials so as to limit impurities such as silicon, carbon and sodium and that crystal growth should be performed under high vacuum so as to reduce the degree of oxidation. The growth should also be performed in a high-purity boron nitride crucible, and the transport agents should be carefully selected so as to avoid contamination of the final product. Charged substitutions led to weaker phonon-scattering for C_{As}, C_B, Si_{As} and Te_{As} than did the neutral counterpart. The opposite behaviour was observed for O_{As}.

Table 40. Impurity Concentrations in BAs

Sample	Si (at%)	Na (at%)	Ca (at%)	Te (at%)
s1	0.047	0.013	0.009	0
s2	0.05	0.01	0.012	0
s3	0.431	0.013	0	0.014

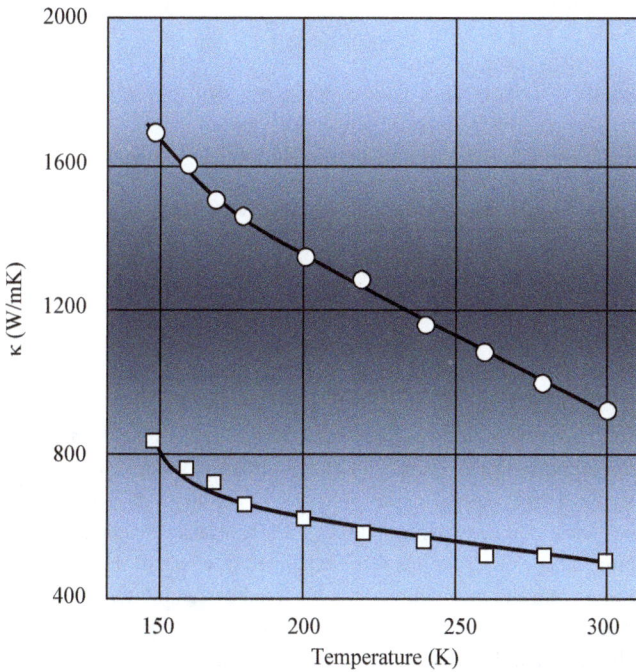

Figure 8. Temperature dependence of the thermal conductivity of Bas
Circles: s1, squares: s2

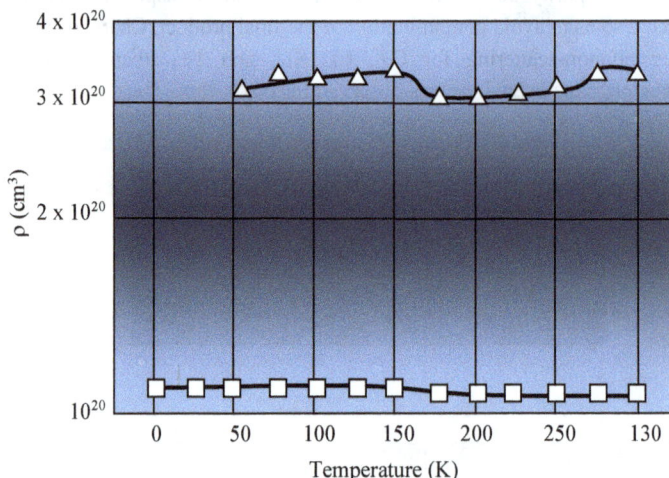

Figure 9. Temperature dependence of the resistivity of Bas
Triangles: s3, squares: s2

Doping thus tends to impair the high thermal conductivity. First-principles calculations were used[147] to determine the reduction in conductivity which was caused by the presence of various group-IV impurities (carbon, silicon, germanium) as a function of their concentration and charge-state. It was discovered that there exists a general trend, in which neutral impurities scatter phonons more strongly than do charged ones. Meanwhile C_B and Ge_{As} impurities exhibit the very weakest phonon-scattering and maintain conductivities of over 1000W/mK; even when present in high densities. The large formation and ionization energies of C_B donors limit their value as n-type dopants. Silicon and germanium produce quite high hole-densities while maintaining a high thermal conductivity. There was an observable fall in conductivity with respect to the charged D-acceptor case upon germanium doping. There was an increase in conductivity upon carbon doping. These observations were explained in terms of the change in orbital occupations in going from the original to the substituted system. Attention was also paid to the peculiarity that, even when the substitution involved a considerable difference in mass, the mass-difference induced scattering could be small. Excessive doping, even to the Fermi-level pinning-point promoted phonon-donor scattering events which could either slow the decrease in thermal conductivity or instead cause it to decrease sharply;

depending upon the type of impurity. It was predicted that Si_{As} and Ge_{As} impurities would be useful p-type dopants, given that both of them led to relatively high free-hole densities while maintaining a conductivity that was far higher than that of any other competing semiconductor.

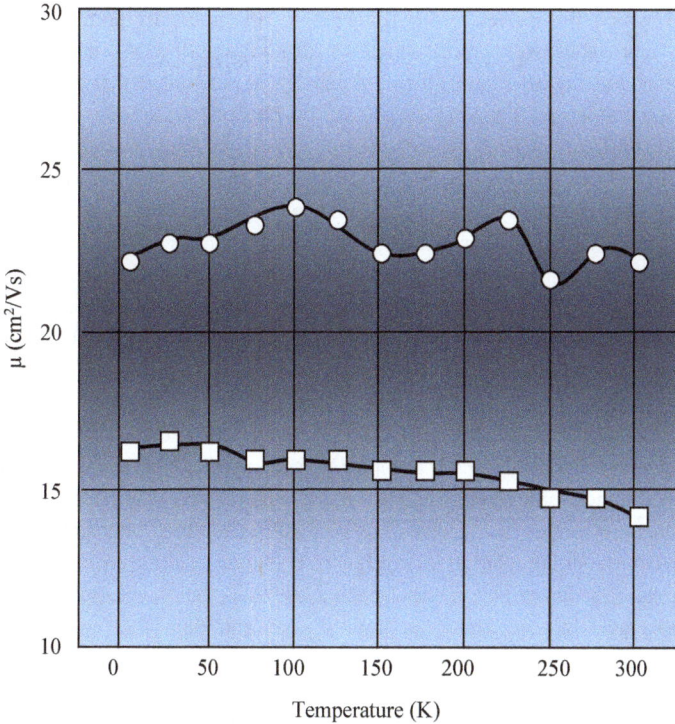

Figure 10. Temperature dependence of the mobility of Bas
Circles: s1, squares: s2

It was noted that substitutional dopants introduced 2 types of perturbation. One type was that of on-site mass perturbations, V_M, which resulted from the mass difference between host and substituted atoms. The other type was that of extended bond perturbations, V_K, which resulted from the distortion of the local bonding environment that was produced by the impurity. The total perturbation, $V = V_M + V_K$, creates a further phonon-scattering which decreases the thermal conductivity. It was assumed that the dopants were randomly distributed throughout the crystal, and that their concentrations were low enough to permit each defect to be treated as an independent scattering center. The phonon thermal transport in the presence of intrinsic 3- and 4-phonon scattering and carbon, silicon and germanium dopants, was deduced by solving the phonon Boltzmann transport equation. The phonon-defect scattering was described by using a T-matrix approach in which the defects were analyzed to all orders of perturbation theory. Theoretical approaches were constructed for determining the V_M and V_K phonon-scattering rates and for calculating the conductivity. It was noted that transport in this material is anomalous in that it is dominated by phonons which are confined to the narrow range of frequencies lying between 4 and 8THz. This was due to the large frequency-gap between the acoustic and optic phonons, and due to a narrow optic-phonon bandwidth. In addition, there was a grouping of acoustic phonon branches with especially weak 4-phonon scattering. This confluence of peculiar features gave rise to particularly large contributions being made to the conductivity by acoustic phonons in this range. Upon increasing the silicon, germanium and carbon impurity-densities, the free-hole density increased, saturated and then finally decreased due to compensation which arose from electrons that were ionized from increasingly large concentrations of donor atoms. Maximum hole densities of 2×10^{18}, 1×10^{18} and $1.5 \times 10^{17}/cm^3$ were found for silicon, germanium and carbon, respectively. The lower value found for carbon was attributed mainly to its greater larger (0.09eV) acceptor ionization when compared to the value of 0.03eV for both germanium and silicon. These results were expected for impurity-formation energies under the arsenic-rich growth conditions imposed during BAs synthesis. When donor compensation was ignored, similar results were found for silicon and germanium up to impurity densities of about $10^{19}/cm^3$, thus suggesting that there was no appreciable gain in hole density at impurity concentrations that were a few times $10^{18}cm^3$. This was especially so when a high conductivity was sought. Under arsenic-rich growth conditions, germanium-doping offered a clear advantage over carbon and silicon doping, as donor-compensation began to be significant only at concentrations which approached $10^{19}cm^3$. For all of the dopants, the conductivity was greater than 600W/mK; even at high dopant densities. This ensured that the conductivity remained far above that of silicon (140W/mK) and GaAs (45W/mK). It was finally observed that, while charged C_B substitution in the absence of C_{As} would lead only to a slight reduction in the conductivity at high densities, the use of carbon as an n-type dopant would be limited by extensive C_B formation and high ionization energies. The fact that charged impurities which are iso-electronic with the substituted species scatter phonons more weakly than do their neutral counterparts promises to be useful in the exploitation of BAs.

The high lattice thermal conductivity of semiconducting cubic boron arsenide has naturally motivated studies of the bulk electronic band structure. An early study[148] of the then-unexplored BAs diatomic structure was performed using high-level multi-reference variational first-principles methods. Potential energy curves were constructed for 42 molecular states dissociating into the first 4 asymptotic channels and covering an energy-range of 5.8eV. This revealed a unique morphology and a rich spectroscopic pattern. The ground-state had $^3\Pi$ symmetry while its first excited $^1\Sigma^+$ state lay about 8kcal/mol higher. A general feature of the potential curves was the existence of avoided crossings that questioned the validity of the Born-Oppenheimer approximation.

Scanning tunnelling spectroscopy was later used[149] to determine the electronic structure of as-grown and *in situ* cleaved surfaces of single crystals. The band-gap, as measured at various internal locations of the cleaved surface was about 2.1eV. This was close to the calculated bulk band-gap value of 2.05eV. The band-gap which was measured within some microns from the edges of the cleaved surface decreased to about 1.9eV. This decrease was attributed to tunnelling from an increased concentration of shallow acceptors. Some of the tunnelling peaks within the band-gap were close to the calculated energy levels for bulk lattice defects and substitutional impurities such as C_{As} and Si_{As} near to the edges of the cleaved surface. Some tunnelling peaks in the band-gap were close to the calculated energy levels for As antisite (As_B) defects, antisite pair (As_B-B_{As}) defects, vacancy defects (V_B, V_{As}) and substitutional C and Si donors on the B-site (C_B, Si_B) in the bulk arsenide. Other tunnelling peaks that did not correspond to bulk impurity or lattice-defect states were also observed within the band-gap.

According to Heyd–Scuseria–Ernzerhof calculations for the relaxed (110) surface these tunnelling peaks can contain contributions arising from intrinsic surface states within the indirect bulk band-gap. Like the tunnelling spectrum obtained from the as-grown surface, the onset of tunnelling from states near to the valence-band maximum occurred at about -0.19eV from the Fermi level within a few microns of the two edges of the cleaved surface; resulting in an apparent band-gap that was close to 1.9eV. This value was about −0.43eV at three interior locations of the cleaved surface and resulted in an apparent band-gap value which was close to 2.1eV. This variation was attributed to a higher concentration of impurities and defects that acted as shallow acceptors near to the two edges. At all of the locations, the measured Fermi level was consistent with p-type conductivity. Previous theoretical studies had shown that carbon and silicon impurities prefer to occupy the arsenic site rather than the boron site, thus leading to p-type conductivity. The calculation used the Heyd–Scuseria–Ernzerhof hybrid functional together with the projector augmented wave method. A mixing parameter of 0.28 yielded an indirect band-gap of 2.05eV for bulk material. The impurity doping energy levels for bulk material were calculated (figure 11). Also calculated were the properties of the (110) surface. For a relaxed surface, calculations indicated boron dangling bond states in the bulk band-gap; about 1.0eV above the bulk valence band maximum. The calculations showed the boron atoms buckled inwards while arsenic atoms buckled outwards. The band-gap value which was measured at the 3 interior locations was some 0.05eV higher than the calculated bulk band-gap, but this discrepancy was within the error caused by effects such as tip-induced band-bending. Such band-bending could increase the measured band-gap value when the impurity dopant concentration was low.

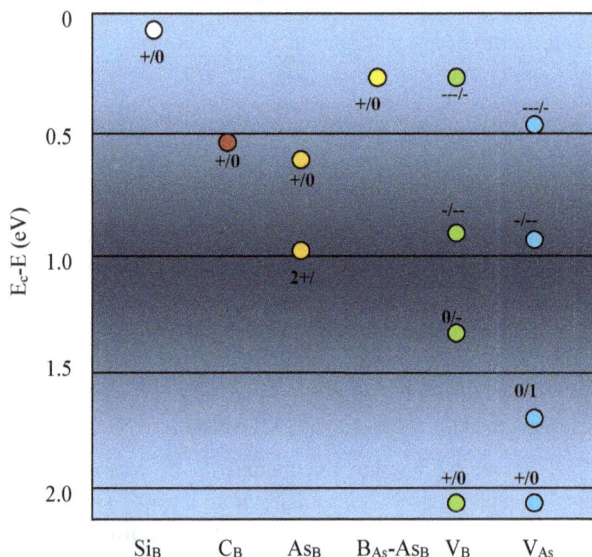

Figure 11. Heyd–Scuseria–Ernzerhof results for defect levels in BAs. The labels indicate the charge states which are involved in the thermodynamic transition level of the defect

It is found[150] that, with 4-phonon scattering included, the thermal conductivity of wurtzite BAs attains values as high as 1036W/mK along the a-b plane at room temperature. This was a decrease of 43% when compared with the results of calculations which were made without considering 4-phonon scattering (figure 12). The similar phonon transport properties of cubic and wurtzite BAs were explained in terms of the similar projected density of states and scattering rates, which in turn arose from the resemblance of the crystal structures. The moment tensor potential of machine-learning approaches proved to be a reliable and efficient method for obtaining high-order interatomic force-constants. Before investigating phonon transport in wurtzite BAs its structural stability was determined by performing *ab initio* molecular dynamics simulations. Energy, temperature and pair-correlation functions for the cubic and wurtzite phases were deduced over 1ps. The pair correlation functions for the phases were very close at a certain temperature because the wurtzite phase had a different stacking sequence of the close-packed planes and could be regarded as being a stacking-fault in the cubic phase. The peak positions of the pair correlation functions of wurtzite BAs, reflecting the structural characteristics, did not vary between 50 and 1100K; thus implying

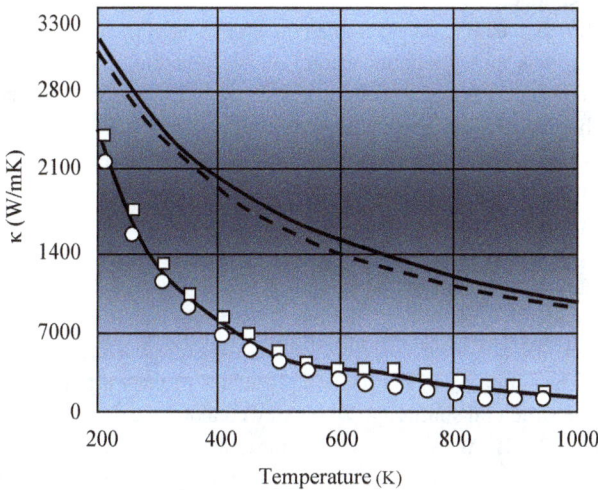

Figure 12. A comparison of the predicted conductivities of cubic BAs including 3-phonon and 4-phonon scattering. Solid curve: density functional theory 3-phonon, broken curve: machine learning potential 3-phonon, squares: density functional theory 3-phonon & 4-phonon, circles: machine learning potential 3-phonon & 4-phonon

structural stability of the material. Because cubic and wurtzite BAs have a comparable energy per atom, a stacking-fault can easily form in the cubic phase, leading to transformation into the wurtzite phase. The Gibbs free energy of the wurtzite phase is almost the same as that of the cubic phase over a wide pressure-range. By using the interatomic force constants arising from the moment tensor potential, an investigation was made of the phonon transport properties of the wurtzite. In the case of this phase there was a phonon frequency gap of 8.5THz; comparable to that of the cubic phase.

Unlike the latter phase, a mixture of acoustic and optical branches below the frequency gap appeared in the wurtzite. Four-phonon scattering was significant for the thermal transport properties of the latter over the entire temperature range. Upon including 4-phonon scattering, the conductivity along the a–b plane decreased from 1808 to 1036W/mK, but it decreased from 1302 to 836W/mK along the c-axis at room temperature. The conductivity of the wurtzite phase along the a–b plane at room temperature was only 14% smaller than that of the cubic phase. Upon including 4-phonon scattering, the conductivity of the wurtzite along the a-axis and that of the cubic phase at room temperature decreased by 43 and 50%, respectively. In a similar manner to the cubic phase, the conductivity of the wurtzite was dominated by phonons with frequencies ranging from 3.0 to 9.0THz, where 3-phonon scattering rates exhibited a dip. Four-phonon scattering was thus a notable feature of the anharmonic effect in the wurtzite. Both the 3-phonon and 4-phonon scattering rates of the wurtzite were very similar to those of the cubic phase. Four-phonon scattering could therefore be expected to lead to the conductivity decreasing as markedly as it did in the cubic material. It seems to be anomalous that both phases have close phonon scattering rates and thermal conductivities but a distinctly different phonon dispersion. In addition to the phonon scattering rate, the group velocity also governs the conductivity. The group velocities of the two phases are quite similar below the phonon frequency gap, and that resemblances leads to the small difference in conductivities. When combined with the similar group velocities, a similarity between the projected density of states suggests that the phonon dispersions may be related. The similarity of the phonon properties of the wurtzite and cubic phases is attributed to the structural similarity. When viewed along the a-axis, both phases are 4-fold coordinated. The B–As bond length in the cubic phase is 2.09Å, and 'changes' anisotropically to 2.09Å in the wurtzite phase. Although the atoms change positions slightly during the cubic ↔ wurtzite martensitic transformation, some nearest-neighbours appear to exchange positions; again limiting any difference between the vibrational behaviours. The interatomic potential is not really changed in going from cubic to wurtzite, resulting in similar phonon properties; including dispersion and scattering. It is also suggested that the phonon dispersions of the structures are related by Brillouin-zone folding.

The ill-effects of phonon scattering can be appreciated by considering a material which is at the other end of the conductivity from BAs. Local cationic off-centering within the overall rock-salt structure of $AgSbSe_2$ was studied[151] in order to explain its very low thermal conductivity of about 0.4W/mK at 300K. The cations were locally off-centered along the crystallographic direction by about 0.2Å. The phonon dispersion which was deduced using density functional theory exhibited weak instabilities that caused local off-center distortions within an anharmonic double-well potential. The local distortion arose from stereochemically-active $5s^2$ lone pairs of antimony. This showed how the local structure influences phonon transport. Although $AgSbSe_2$ maintained an overall cubic structure, the local environment remained distorted along the <100> directions. As noted above, local off-centering is a result of the stereochemical activity of $5s^2$ lone pairs of Sb^{3+}. This exerted a repulsive effect upon the Sb-Se bonds and thereby distorted the $SbSe_6$ octahedron. The lone-pair-induced off-centering also affected the neighbouring $AgSe_6$ octahedra. The cation off-centering results in a local bonding heterogeneity involving shorter and longer M-Se bond pairs. The overall structure of $AgSbSe_2$ could be appreciated by combining a local <100> distorted model at r < 4.5Å and an overall cubic model at 4.5Å < r < 20Å. The close agreement of local and overall peaks reflected the local displacement of cations within the average cubic sub-lattice. Weak instability in the phonon-dispersion supports the local off-centering distortions within an anharmonic double-well potential. Raman spectroscopy confirmed the presence of a locally distorted asymmetry by the presence of low-energy Raman inactive modes. The local distortion produced appreciable lattice strain and this, combined with a high anharmonicity and low-energy optical phonon modes, increased the phonon scattering which in turn finally led to an intrinsically ultra-low lattice thermal conductivity.

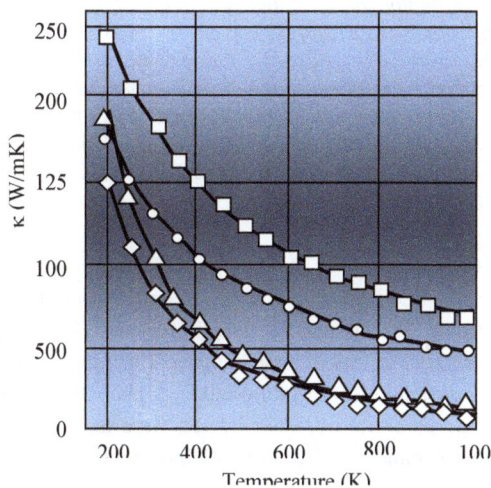

Figure 13. Thermal conductivity of wurtzite BAs, including 3-phonon scattering. Squares: a-axis 3-phonon, circles: c-axis 3-phonon, triangles: a-axis 3-phonon & 4-phonon, diamonds: c-axis 3-phonon & 4-phonon

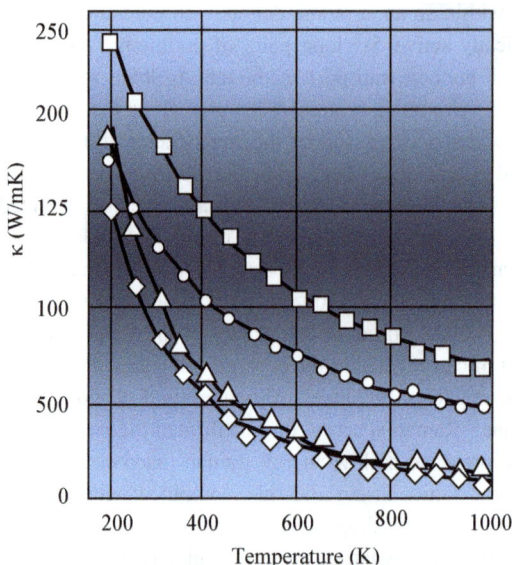

Figure 14. A comparison of conductivities predicted using different interatomic force constants for 0K and other temperatures. Top curve: a-axis 3-phonon 0K, second curve: c-axis 3-phonon 0K, third curve: a-axis 3-phonon & 4-phonon 0K, Lowest curve: c-axis 3-phonon & 4-phonon 0K, Squares: a-axis 3-phonon, temperature-dependent effective potential, circles: c-axis 3-phonon, temperature-dependent effective potential, triangles: a-axis 3-phonon & 4-phonon with temperature-dependent effective potential, diamonds: c-axis: 3-phonon & 4-phonon with temperature-dependent effective potential

An important observation is that it may be easier to synthesize wurtzite-phase nanostructures. The renormalized conductivity of the wurtzite at room temperature is a function of size (figure 13). When the length reaches 100nm, the conductivity in nanostructures is some 10% of that of bulk samples. This size effect is much greater than that found in other systems. Temperature-induced phonon renormalization can also significantly affect the conductivity. Calculations were made of the interatomic force constants at finite temperatures and used to obtain the temperature-dependent phonon dispersions and the conductivity of the wurtzite phase. The effect of phonon-softening could not be neglected at high temperatures. Without including 4-phonon scattering, the conductivity did not change appreciably when compared to the conductivity which resulted from using interatomic force constants at 0K; especially along the c-axis. Upon including 4-phonon scattering, the effect of phonon renormalization upon the conductivity became non-negligible. In particular the difference in conductivity, with and without the temperature effect upon the interatomic force constants, increased up to 20% above 600K. This suggested that the temperature effect weakens the 4-phonon scattering in the wurtzite phase but barely affects the 3-phonon scattering.

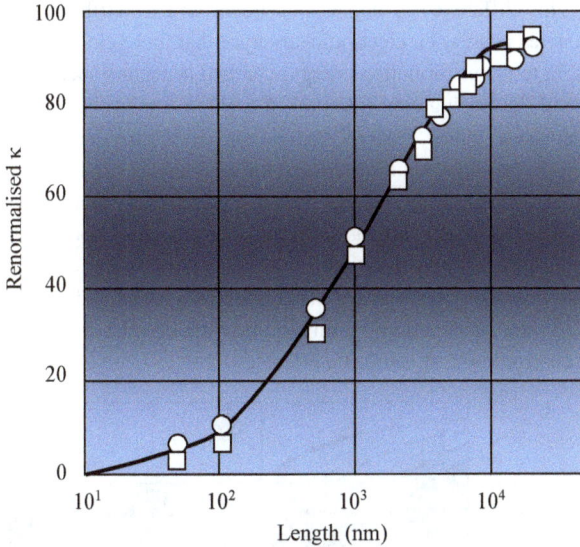

Figure 15. Renormalized conductivity of wurtzite Bas
at room-temperature as a function of size, Squares: a-axis, circles: c-axis

Theoretical methods have been developed in order to aid the understanding of the physical properties of cubic BAs. A common problem[152] is that the conductivity varies widely from point to point within the same sample: the room-temperature value can range from less than 500 to more than 1200W/mK. These variations indicate the existence of defects. The reactions that occur in quartz tubes during chemical vapour transport growth of BAs are also unknown. The effects of impurities have been studied by using first-principles methods. A reduction in thermal conductivity, due to phonon scattering, occurs upon doping with carbon, silicon or germanium (figure b). The mass-difference between the dopant and host atoms results in mass-defect scattering and thus affects the conductivity. It is noted that C_{As} and Ge_B defects have the greatest tendency to decrease the conductivity. The Ge_{As} and C_B defects cause relatively smaller decreases. The motion of heavy atoms such as arsenic predominates in almost all of the acoustic phonon modes through the Brillouin zone, and the presence of mass defects at arsenic sites leads to a stronger scattering of acoustic phonons than that at the lighter boron sites. Neutral dopants reduce the conductivity more than do their charged equivalents for a given dopant density. This difference is dopant-dependent and is attributed to the fact that ionized charged states possess an electronic structure that more closely resembles that of the original host. In terms of formation energy, charged acceptors such as Si_{As}^-, Ge_{As}^- and C_{As}^- form first. With increasing concentration of a given dopant the formation energy of the acceptor increases, while that of the donor decreases; so the donor can be compensated. The Fermi-level pinning values for arsenic-rich or boron-rich conditions vary for different dopants. There are changes in the curves for germanium and carbon dopants, due to the effects of compensation, and the changes depend upon the dopant concentration (figure 16). When BAs is grown under boron-rich conditions, the p-type germanium dopant does not greatly affect the conductivity; even at dopant concentrations of the order of $10^{19}/cm^3$. Compensation only slightly affects carbon-doping, and silicon-doping is hardly changed by compensation effects.

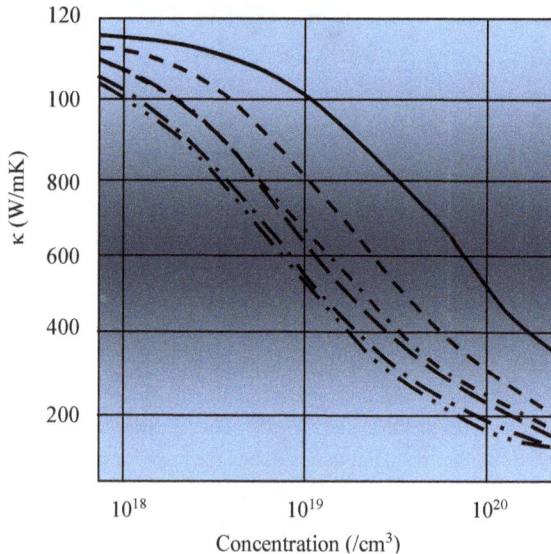

Figure 16. Lattice thermal conductivity of BAs as a function of dopant concentration at 300K for each of the considered impurities. Solid line: Ge_{As}, short-dash line: C_B, short-

Materials Research Forum LLC
https://doi.org/10.21741/9781644902233

dash dot line: Si_B, long-dash line: Si_{As}, long-dash dot line: C_{As}, long-dash dot dot line: Ge_B

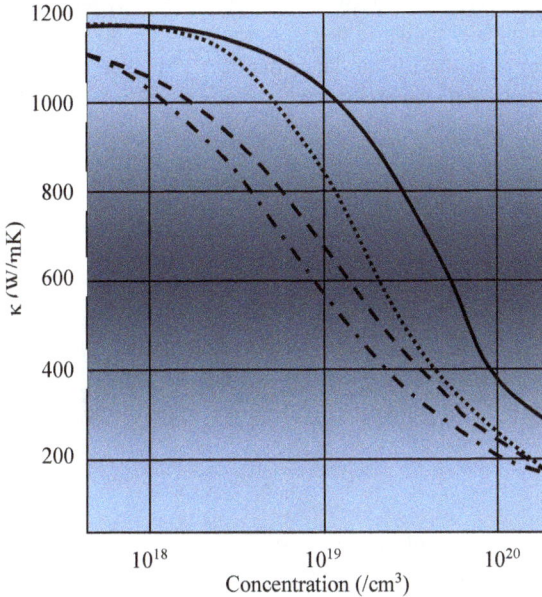

Figure 17. Lattice thermal conductivity of BAs as a function of germanium- and carbon-impurity concentrations at 300K Solid line: Ge, B-rich, dotted line: Ge, As-rich, dash line: C, As-rich, dash dot line: C, B-rich

The III-V boron compounds exhibit abnormal thermal conductivities, but the thermal conductivity of boron arsenide is much higher than that of boron phosphide or boron antimonide. First-principles calculations and the Boltzmann transport equation were used[153] to study the conductivity properties of the III-V boron compounds. A comparison of the IV and III-V semiconductors showed that the high thermal conductivity of boron arsenide is due mainly to the existence of a large frequency-gap between the acoustic and optical branches (table 41). The energy sum of two acoustic phonons is less than the energy of one optical phonon, and cannot meet the energy-conservation requirements of 3-phonon scattering. This greatly restricts the possibility of scattering of 3 phonons. This was distinctly different to the high thermal conductivity of diamond, which is due mainly to its high acoustic phonon group-velocity. Meanwhile boron phosphide also has a relatively large acoustic phonon group-velocity but the frequency gap is relatively small and cannot effectively suppress 3-phonon scattering. The thermal conductivity of boron phosphide is therefore less than that of boron arsenide. Finally, although the frequency-gap of boron antimonide is similar to that of boron arsenide, the thermal conductivity of the former is lower than that of the latter due to its smaller acoustic-phonon group-velocity and larger coupling matrix element.

Table 41. Maximum vibration frequency of the transverse optical mode, maximum vibration frequency of the transverse acoustic mode and frequency-gap of boron compounds

Property	BAs	BP	BSb
ω_{TO} (THz)	19.19	22.60	16.90
ω_{TA} (THz)	8.93	15.15	6.48
Δ (THz)	10.26	7.45	10.42

A detailed study was made[154] of the growth of single crystals of zincblende BAs having various isotopic ratios. This revealed that the thermal conductivity of isotopically pure material BAs was at least 10% greater than that of material which was prepared using natural boron (table 42). Raman spectroscopy revealed differences in scattering among the various samples. There was an obvious change in the phonon frequency of isotopically pure BAs, thus indicating that boron isotopes affected the phonon interactions. Time-domain thermoreflectance methods were used to measure the conductivity because traditional heat-transport measurements cannot be used, due to near-zero temperature difference existing over short sample lengths. As indicated in the table, the conductivity was measured at many points and the highest and lowest values were noted. The highest room-temperature conductivity of ^{10}BAs (1260W/mK) was higher than that (1160W/mK) of ^{nat}BAs. The highest room-temperature conductivity of ^{11}BAs (1180W/mK) was lower than that of ^{10}BAs. Single crystals were also grown by hand-mixing boron isotopes in various ratios. The highest conductivities found for as-prepared $^{10}B_{0.2}{}^{11}B_{0.8}$As and $^{10}B_{0.5}{}^{11}B_{0.5}$As were less than 300W/mK. These values were much lower than expected, and this was attributed to an effect of the manual mixing.

The Boron Arsenides   Materials Research Forum LLC
Materials Research Foundations **138** (2023)   https://doi.org/10.21741/9781644902233

Table 42. Thermal conductivity of isotopically pure BAs

Composition	κ_{max} (W/mK)	κ_{min} (W/mK)
^{10}BAs	1260	689
^{11}BAs	1180	381
^{nat}BAs	1160	640

Recent experimental data have confirmed[155] the predicted high thermal conductivity and have revealed large conductivity variations which were associated with extended and point defects. The peak conductivity furnished valuable information on the competition between intrinsic phonon-phonon scattering processes and extrinsic boundary and defect scattering processes. The peak measured conductivity appeared at temperatures of between 120 and 150K, and varied from 410 to 830W/mK. The measured thermal conductivity agreed with theoretical predictions over the entire temperature range, and this suggested that the boundary-scattering mean free path was about 4μm in two of the samples and 5μm in a third. The variation in the magnitude of the peak conductivity disclosed an order-of-magnitude difference in the strength of point-defect scattering. The phonon-defect scattering behaviour was closely related to the measured electronic Raman scattering background, to the impurity concentrations revealed by secondary ion mass spectroscopy and to the Hall concentration and mobility of p-type samples; apart from an anomalously high hole concentration which appeared in one of the samples. This indicated the existence of a non-uniform impurity distribution.

Two new optical techniques, photoluminescence-mapping and time-domain thermo-photoluminescence, have been developed[156] for making rapid non-destructive measurements of thermal conductivity without requiring extensive sample preparation. The photoluminescence -mapping method provides almost real-time imaging of crystal quality and conductivity over mm-sized crystal surfaces. The time-domain thermo-photoluminescence method permits the choice of any spot on the sample surface for the purpose of measuring the local conductivity by using nanosecond laser pulses. These techniques reveal that supposedly single crystals are in fact non-uniform with regard to conductivity and, moreover, are composed of domains which possess quite different conductivities. Because both methods are based upon the band-edge photoluminescence, and its dependence upon temperature, they can be applied to other semiconductors.

The next obvious question after considering the effect of impurities is to ask how an additional element will change the properties. A careful investigation was made[157] of the fundamental lattice-vibrational spectra in ternary compounds and their thermal conductivity by using a predictive first-principles approach. The phonon transport in B-X-C groups, where X was nitrogen, phosphorus or arsenic, was quantified for various crystal structures and high-order anharmonicity involving a 4-phonon process. The results revealed an ultra-high room-temperature thermal conductivity of up to 2100W/mK that was due to strong carbon-carbon bonding. Such conductivities were even beyond that of boron arsenide. This offered fundamental insights into the atomistic causes of thermal conductivity.

Thermodynamic data on the enthalpy, entropy and Gibbs free energy of BAs were deduced[158] from experimental measurements of its heat capacity at 298 to 1150K (table 43). The thermodynamic properties were H = -8.6kJ/mol, S = 81.0J/molK and G = -73.4kJ/mol. The formation reaction became endothermic at 984K. Thermodynamic analysis also showed that iodine is the best halogen for obtaining monocrystals via chemical vapour transport. Monocrystals were grown by using various combinations of precursors. One approach offered a low transport-rate, resulting in slow monocrystal growth, but did not require excess arsenic to be present in the tube: excess arsenic could lead to a build-up of internal pressure, The use of shorter tubes could efficiently increase the transport rate. The best-quality crystals were made by using a second approach, and the resultant crystals had thermal conductivities which approached the theoretical prediction; thus indicating a lower defect content. The boron used could be a source of impurities when using this approach. When using the second and third approaches excess arsenic helped to control the transport-rate. The third approach offered a higher transport rate and produced more crystals at the cold end.

This third approach used preselected pure monocrystals of higher thermal conductivity as the source[159]. As noted above, optimization of the crystal quality was impaired by the dubious purity of the boron source and by contamination of the quartz tubes. It was found that, compared to methods which produced good crystals from high-purity boron, the use of BAs crystals as the source material could lead to further improvements.

Table 43. Properties of BAs at 298 to 1150K

Temperature (K)	C_P (J/molK)	S (J/molK)	G (kJ/mol)	H (kJ/mol)
298.15	32.382	40.00	−41.93	−30.00
300	32.561	40.20	−42.00	−29.94
350	36.592	45.54	−44.14	−28.21
400	39.439	50.62	−46.55	−26.30
450	41.479	55.39	−49.20	−24.27
500	42.954	59.84	−52.08	−22.16
550	44.026	63.99	−55.18	−19.99
600	44.809	67.86	−58.48	−17.76
650	45.387	71.47	−61.96	−15.51
700	45.819	74.85	−65.62	−13.23
750	46.155	78.02	−69.44	−10.93
800	46.429	81.01	−73.42	−8.61
850	46.671	83.83	−77.54	−6.29
900	46.905	86.51	−81.80	−3.95
950	47.149	89.05	−86.19	−1.60
1000	47.418	91.47	−90.70	0.77
1050	47.726	93.79	−95.34	3.15
1100	48.083	96.02	−100.08	5.54
1150	48.498	98.17	−104.94	7.96

Heat-spreading

Attempts have been made to create passive ultra-conductive thermal metamaterials which consist of nothing but bulk natural materials. By means of local thermal resistance regulation by a vertical thermal transport channel, the thermal metamaterials offered an effective conductivity of 1915W/mK by using only naturally occurring materials. The purely conductive modulation, without any external energy being used, was comparable to active counterparts[160]. The control of heat transfer is fundamental to thermal management but heat transport by static material via thermal conduction is limited by the thermal conductivities of natural materials, which range from 0.023 W/mK for air to 429 W/mK for silver. Ultra-conductivity can be achieved in effect via thermal enhancement by phase-changes or heat sources. Emerging metamaterials, based upon digital conductivity distributions are making significant advances in guiding heat fluxes in elaborate arrangements The new discovery of advective metamaterials has further introduced conductive and inhomogeneous tunability. The use of external energy is assumed to be necessary for inducing ultra-conductive heat transfer. It is therefore counter-intuitive that the thermal-transport limitations of natural materials can be overcome by pure conduction. But considering passive thermal material, a simple configuration and close fitting to complex morphologies may offer a perfect replacement for active alternatives if ultra-conductive transport is achieved within the constraints of pure conduction. Inspired by thermal resistance, an extended surface area produces considerable enhancement of the transport process. An extra transport channel to an interphase could be a useful means for overcoming conductive limitations. By analogy, interphases could be used to create efficient thermal management by tunable thermal interfacial effects. A mechanism for creating ultra-conductive thermal metamaterials was proposed which consisted of nothing but ordinary materials. Ultra-conductive thermal metamaterials can achieve effective thermal conductivity.

The great promise of BAs is due to the fact that its high thermal conductivity will aid the cooling of electronic devices. Attention is therefore paid as to how to optimize this process. It was shown that BAs cooling substrates can be heterogeneously integrated with metals, a wide-bandgap semiconductor such as gallium nitride and high electron-mobility transistor devices. It was found[161] that GaN/BAs structures exhibited a high thermal boundary conductance of 250 MW/m²K. A comparison of device-level hot-spot temperatures with length-dependent scaling from 100μm to 100nm showed that the cooling performance of BAs exceeded that of current diamond coatings. High electron-mobility AlGaN/GaN transistors, with BAs substrates, had much lower hot-spot temperatures than did diamond and silicon carbide for the same transistor power-density. The good thermal-management performance of BAs was attributed to both the unique phonon band-structures and to interface-matching. A numerical investigation has been made[162] of the optimum distribution of a limited amount of high thermal conductivity material so as to enhance heat-removal from 3-dimensional integrated circuits. The structure of the heat-spreader was designed to be a composite of a high thermal conductivity material (boron arsenide) and moderate thermal conductivity material (copper). The volume-ratio of high-conductivity inserts within the overall volume of the spreader was set at a suitable fixed ratio. Two boundary-conditions, constant temperature and variable temperature, were considered. In order to assess the result of adding high-conductivity inserts, with regard to the cooling effect of the heat spreader, divers patterns of single-ring and double-ring inserts were used. The optimum location of the rings was determined, as well as the optimum distribution of high-conductivity material between the inner and outer rings. It was found that the maximum temperature of the 3-dimensional integrated circuit was reduced by up to 10%, while the size of the heat-sink and heat-spreader were simultaneously decreased by up to 200%. In a related study[163] the inserts were distributed in 3 main configurations: radial, one level of pairing and two levels of pairing. The heat-spreader here consisted of a composite of copper and blades made of boron arsenide. The structures which corresponded to the lowest maximum temperature of the device, given a fixed ratio of the boron arsenide volume to the overall heat-spreader volume were determined. Four different boundary conditions were assumed here. It was found that, for a constant temperature, a variable temperature and convection heat-transfer boundary conditions, the maximum temperature of the entire structure could be reduced by 13.7, 11.9 and 13.9%, respectively. The size of the plain heat-spreader was 200% larger than that of the non-embedded one.

The thermal boundary conductance is another important factor for heat dissipation in light-emitting diodes. The boundary conductance between high-conductivity boron arsenide and silicon was predicted[164] by means of non-equilibrium molecular dynamics simulations. From the thermal conductivity accumulation function with respect to phonon frequency, it was deduced that the predominant phonon frequencies for heat conduction in BAs are very different to those in silicon. The non-equilibrium molecular dynamics simulations nevertheless indicated that the thermal boundary conductance of the BAs/Si interface was high when compared to most other interfaces. This was true even though there was a considerable frequency-mismatch in the thermal conductivity accumulation function between BAs and silicon. The main reason for the high thermal boundary conductance was the overlap of the phonon density of states between BAs and silicon at frequencies of 5 to 8THz. The range of the predicted thermal boundary conductance of the BAs/Si interface was 200 to 300MW/m^{2}K at 300 to 700K. The thermal boundary conductances of Si/BAs and Si/Ge interfaces were close to one other. A useful method for measuring the thermal conductivity of materials and the thermal conductance of their interfaces is that of time-domain thermoreflectance. Its utility was demonstrated[165] by citing studies of the high thermal conductivities of crystals of cubic boron phosphide and boron arsenide.

A novel high-performance thermal interface was developed[166] which went beyond the existing state-of-the-art and was upon self-assembled cubic boron arsenide. This arsenide exhibited thermal conductivities of up to 21W/mK, with an elastic compliance which was similar to that of soft biological tissue, down to 100kPa, and which was achieved by incorporating arsenide microcrystals into a polymer. The composite also exhibited a high flexibility and retained a high conductivity over after at least 500 bending cycles. When integrated with a power LED, the cooling performance of the self-assembled arsenide went beyond the current state-of-the-art limit by an up to 45C reduction in the hot-spot temperature.

About the Author

Dr. Fisher has wide knowledge and experience of the fields of engineering, metallurgy and solid-state physics, beginning with work at Rolls-Royce Aero Engines on turbine-blade research, related to the Concord supersonic passenger-aircraft project, which led to a BSc degree (1971) from the University of Wales. This was followed by theoretical and experimental work on the directional solidification of eutectic alloys having the ultimate aim of developing composite turbine blades. This work led to a doctoral degree (1978) from the Swiss Federal Institute of Technology (Lausanne). He then acted for many years as an editor of various academic journals, in particular *Defect and Diffusion Forum*. In recent years he has specialized in writing monographs which introduce readers to the most rapidly developing ideas in the fields of engineering, metallurgy and solid-state physics. He is co-author of the widely-cited student textbook, *Fundamentals of Solidification*, a new (5th fully-revised) edition of which is soon to appear. Google Scholar credits him with 8687 citations and a lifetime h-index of 14.

References

[1] Slack, G.A., Journal of Physics and Chemistry of Solids 34, 1973, 321-335. https://doi.org/10.1016/0022-3697(73)90092-9

[2] Li, S., Zheng, Q., Lv, Y., Liu, X., Wang, X., Huang, P.Y., Cahill, D.G., Lv, B., Science, 361[6402] 2018, 579-581. https://doi.org/10.1126/science.aat8982

[3] Tian, F., Song, B., Chen, X., Ravichandran, N.K., Lv, Y., Chen, K., Sullivan, S., Kim, J., Zhou, Y., Liu, T.H., Goni, M., Ding, Z., Sun, J., Gamage, G.A.G.U., Sun, H., Ziyaee, H., Huyan, S., Deng, L., Zhou, J., Schmidt, A.J., Chen, S., Chu, C.W., Huang, P.Y., Broido, D., Shi, L., Chen, G., Ren, Z., Science, 361[6402] 2018, 582-585. https://doi.org/10.1126/science.aat7932

[4] Liu, J., Zhao, Y., Journal of Synthetic Crystals, 50[2] 2021, 391-396.

[5] Ownby, P.D., Journal of the American Ceramic Society, 58[7-8] 1975, 359-360. https://doi.org/10.1111/j.1151-2916.1975.tb11514.x

[6] Perri, J.A., LaPlaca, S., Post, B., Acta Crystalografica, 11, 1958, 310. https://doi.org/10.1107/S0365110X58000827

[7] Williams, F.V., Canadian Patent CA623777.

[8] Xia, Y., Hegde, V.I., Pal, K., Hua, X., Gaines, D., Patel, S., He, J., Aykol, M., Wolverton, C., Physical Review X, 10[4] 2020, 041029. https://doi.org/10.1103/PhysRevX.10.041029

[9] Yang, X., Feng, T., Kang, J.S., Hu, Y., Li, J., Ruan, X., Physical Review B, 101[16] 2020, 161202. https://doi.org/10.1103/PhysRevB.101.161202

[10] Gamage, G.A., Chen, K., Chen, G., Tian, F., Ren, Z., Materials Today Physics, 11, 2019, 100160. https://doi.org/10.1016/j.mtphys.2019.100160

[11] Hirayama, M., Shohno, K., Japanese Journal of Applied Physics - I, 12[12] 1973, 1960-1961. https://doi.org/10.1143/JJAP.12.1960

[12] Beckel, C.L., Lu, N., Abbott, B., Yousaf, M., Inorganica Chimica Acta, 289[1-2] 1999, 198-209. https://doi.org/10.1016/S0020-1693(99)00069-9

[13] Cherednichenko, K.A., Le Godec, Y., Solozhenko, V.L., High Pressure Research, 38[3] 2018, 224-231. https://doi.org/10.1080/08957959.2018.1476507

[14] Daoud, S., Bioud, N., Bouarissa, N., Materials Science in Semiconductor Processing, 31, 2015, 124-130. https://doi.org/10.1016/j.mssp.2014.11.024

[15] Zhou, Y., Dong, Z.Y., Hsieh, W.P., Goncharov, A.F., Chen, X.J., Nature Reviews

Physics, 4[5] 2022, 319-335. https://doi.org/10.1038/s42254-022-00423-9

[16] Wang, L., Tian, F., Liang, X., Fu, Y., Mu, X., Sun, J., Zhou, X.F., Luo, K., Zhang, Y., Zhao, Z., Xu, B., Ren, Z., Gao, G., Physical Review B, 99[17] 2019, 174104. https://doi.org/10.1103/PhysRevB.99.174104

[17] Fan, Z., Wang, B., Xu, X., Cao, X., Wang, Y., Physica Status Solidi B, 248[5] 2011, 1242-1247. https://doi.org/10.1002/pssb.201046406

[18] Gray, J.C., Pomeroy, J.W., Kuball, M., Xu, Z., Edgar, J.H., Chen, H., Dudley, M., Journal of Applied Physics, 103[9] 2008, 093537. https://doi.org/10.1063/1.2919785

[19] Zhang, C., Zeng, Z., Sun, Q., Chen, Y., Journal of Applied Physics, 130[20] 2021, 205101. https://doi.org/10.1063/5.0065889

[20] Liu, W., Wu, Y., Hong, Y., Zhang, Z., Yue, Y., Zhang, J., Nanotechnology, 33[16] 2022, 162501. https://doi.org/10.1088/1361-6528/ac46d7

[21] Rodriguez, A., Liu, Y., Hu, M., Physical Review B, 102[3] 2020, 035203. https://doi.org/10.1103/PhysRevB.102.035203

[22] Ravichandran, N.K., Broido, D., Physical Review X, 10[2] 2020, 021063. https://doi.org/10.1103/PhysRevX.10.021063

[23] Gul, R., Cui, Y., Bolotnikov, A.E., Camarda, G.S., Egarievwe, S.U., Hossain, A., Roy, U.N., Yang, G., Edgar, J.H., Nwagwu, U., James, R.B., AIP Advances, 6[2] 2016, 025206. https://doi.org/10.1063/1.4941937

[24] Whiteley, C.E., Kirkham, M.J., Edgar, J.H., Journal of Physics and Chemistry of Solids, 74[5] 2013, 673-676. https://doi.org/10.1016/j.jpcs.2012.12.026

[25] Frye, C.D., Edgar, J.H., Zhang, Y., Cooper, K., Nyakiti, L.O., Gaskill, D.K., Materials Research Society Symposium Proceedings, 1439, 2012, 69-75. https://doi.org/10.1557/opl.2012.1156

[26] Whiteley, C.E., Zhang, Y., Gong, Y., Bakalova, S., Mayo, A., Edgar, J.H., Kuball, M., Journal of Crystal Growth, 318[1] 2011, 553-557. https://doi.org/10.1016/j.jcrysgro.2010.10.057

[27] Whiteley, C.E., Zhang, Y., Mayo, A., Edgar, J.H., Gong, Y., Kuball, M., Dudley, M., Materials Research Society Symposium Proceedings, 1307, 2011, 66-72. https://doi.org/10.1557/opl.2011.502

[28] Whiteley, C.E., Mayo, A., Edgar, J.H., Dudley, M., Zhang, Y., Materials Research Society Symposium Proceedings, 1307, 2011, 79-85.

https://doi.org/10.1557/opl.2011.504

[29] Yu, Z., Hui, C., Dudley, M., Yi, Z., Edgar, J.H., Gong, Y., Bakalova, S., Kuball, M., Zhang, L., Su, D., Kisslinger, K., Zhu, Y., Materials Research Society Symposium Proceedings, 1246, 2010, 71-76.

[30] Wang, R.H., Zubia, D., O'Neil, T., Emin, D., Aselage, T., Zhang, W., Hersee, S.D., Journal of Electronic Materials, 29[11] 2000, 1304-1306. https://doi.org/10.1007/s11664-000-0129-x

[31] Nagarajan, R., Edgar, J.H., Pomeroy, J., Kuball, M., Aselage, T., Materials Research Society Symposium - Proceedings, 764, 2003, 283-287. https://doi.org/10.1557/PROC-764-C3.58

[32] Vetter, W.M., Nagarajan, R., Edgar, J.H., Dudley, M., Materials Letters, 58[7-8] 2004, 1331-1335. https://doi.org/10.1016/j.matlet.2003.09.042

[33] Zhang, Y., Chen, H., Dudley, M., Zhang, Y., Edgar, J.H., Gong, Y., Bakalova, S., Kuball, M., Zhang, L., Su, D., Zhu, Y., Materials Research Society Symposium Proceedings, 1307, 2011, 15-20. https://doi.org/10.1557/opl.2011.316

[34] Gong, Y., Zhang, Y., Dudley, M., Zhang, Y., Edgar, J.H., Heard, P.J., Kuball, M., Journal of Applied Physics, 108[8] 2010, 084906. https://doi.org/10.1063/1.3486518

[35] Frye, C.D., Edgar, J.H., Ohkubo, I., Mori, T., Journal of the Physical Society of Japan, 82[9] 2013, 095001. https://doi.org/10.7566/JPSJ.82.095001

[36] Gong, Y., Tapajna, M., Bakalova, S., Zhang, Y., Edgar, J.H., Zhang, Y., Dudley, M., Hopkins, M., Kuball, M., Applied Physics Letters, 96[22] 2010, 223506. https://doi.org/10.1063/1.3443712

[37] Hart, G.L.W., Zunger, A., Physical Review B, 62[20] 2000, 13522-13537. https://doi.org/10.1103/PhysRevB.62.13522

[38] Wu, J., Zhu, H., Hou, D., Ji, C., Whiteley, C.E., Edgar, J.H., Ma, Y., Journal of Physics and Chemistry of Solids, 72[2] 2011, 144-146. https://doi.org/10.1016/j.jpcs.2010.12.005

[39] Foss, C., Aksamija, Z., Nanotechnology, 32[40] 2021, 405206. https://doi.org/10.1088/1361-6528/ac0d7d

[40] Raeisi, M., Ahmadi, S., Rajabpour, A., Nanoscale, 11[45] 2019, 21799-21810. https://doi.org/10.1039/C9NR06283A

[41] Zhuang, H.L., Hennig, R.G., Applied Physics Letters, 101[15] 2012, 153109.

https://doi.org/10.1063/1.4758465

[42] Srivastava, A., Sharma, M., Tyagi, N., Kothari, S.L., Journal of Computational and Theoretical Nanoscience, 9[10] 2012, 1693-1699. https://doi.org/10.1166/jctn.2012.2266

[43] Bhushan, B.S., Jain, S.K., Srivastava, A., Advanced Science Letters, 21[9] 2015, 2850-2854. https://doi.org/10.1166/asl.2015.6389

[44] Dong, Y., Zeng, B., Zhang, X., Li, M., He, J., Long, M., Journal of Applied Physics, 126[12] 2019, 124303. https://doi.org/10.1063/1.5110868

[45] Yu, C., Hu, Y., He, J., Lu, S., Li, D., Chen, J., Applied Physics Letters, 120[13] 2022, 132201. https://doi.org/10.1063/5.0086608

[46] Manoharan, K., Subramanian, V., ACS Omega, 3[8] 2018, 9533-9543. https://doi.org/10.1021/acsomega.8b00946

[47] Freer, R., Ekren, D., Ghosh, T., Biswas, K., Qiu, P., Wan, S., Chen, L., Han, S., Fu, C., Zhu, T., Ashiquzzaman Shawon, A.K.M., Zevalkink, A., Imasato, K., Snyder, G.J., Ozen, M., Saglik, K., Aydemir, U., Cardoso-Gil, R., Svanidze, E., Funahashi, R., Powell, A.V., Mukherjee, S., Tippireddy, S., Vaqueiro, P., Gascoin, F., Kyratsi, T., Sauerschnig, P., Mori, T., Journal of Physics Energy, 4[2] 2022, 022002. https://doi.org/10.1088/2515-7655/ac49dc

[48] Yu, H., Li, Y., Wei, D., Guo, G., Feng, Z., Ye, J., Luo, Q., Ma, Y., Tang, Y., Dai, X., Micro and Nanostructures, 165, 2022, 207188. https://doi.org/10.1016/j.micrna.2022.207188

[49] Ullah, S., Denis, P.A., Menezes, M.G., Sato, F., Applied Surface Science, 493, 2019, 308-319. https://doi.org/10.1016/j.apsusc.2019.07.030

[50] Ullah, S., Denis, P.A., Sato, F., Computational Materials Science, 170, 2019, 109186. https://doi.org/10.1016/j.commatsci.2019.109186

[51] Lalngaihawmi, R., Vanlalruata, B., Hnamte, L., Thapa, R.K., International Conference on Electrical, Electronics, and Optimization Techniques, 2016, 1162-1166.

[52] Ullah, S., Denis, P.A., Sato, F., International Journal of Quantum Chemistry, 119[18] 2019, e25975. https://doi.org/10.1002/qua.25975

[53] Khossossi, N., Banerjee, A., Benhouria, Y., Essaoudi, I., Ainane, A., Ahuja, R., Physical Chemistry Chemical Physics, 21[33] 2019, 18328-18337. https://doi.org/10.1039/C9CP03242H

[54] Brems, M.R., Willatzen, M., New Journal of Physics, 21[9] 2019, 093030. https://doi.org/10.1088/1367-2630/ab3d78

[55] Brems, M.R., Willatzen, M., Proceedings of the International Conference on Numerical Simulation of Optoelectronic Devices, 2019, 115-116.

[56] Alrebdi, T.A., Idrees, M., Munawar, M., Alkallas, F.H., Amin, B., Chemical Physics, 552, 2022, 111374 https://doi.org/10.1016/j.chemphys.2021.111374

[57] Zhu, Q., Tong, Q., Sun, H., Wang, Y., Yao, W., 2D Materials, 6[4] 2019, 045037. https://doi.org/10.1088/2053-1583/ab38d4

[58] Ren, J., Kong, W., Ni, J., Nanoscale Research Letters, 14, 2019, 133. https://doi.org/10.1186/s11671-019-2972-4

[59] Shahriar, R., Hoque, K.S., Tristant, D., Zubair, A., Applied Surface Science, 600, 2022, 154053. https://doi.org/10.1016/j.apsusc.2022.154053

[60] Yu, L., Tian, Y., Zheng, X., Wang, H., Shen, C., Qin, G., International Journal of Thermal Sciences, 174, 2022, 107438. https://doi.org/10.1016/j.ijthermalsci.2021.107438

[61] Bi, S., Chang, Z., Yuan, K., Sun, Z., Zhang, X., Gao, Y., Tang, D., Journal of Applied Physics, 132[11] 2022, 114301. https://doi.org/10.1063/5.0073473

[62] Hu, Y., Yin, Y., Ding, G., Liu, J., Zhou, H., Feng, W., Zhang, G., Li, D., Materials Today Physics, 17, 2021, 100346. https://doi.org/10.1016/j.mtphys.2021.100346

[63] Ouyang, Y., Yu, C., He, J., Jiang, P., Ren, W., Chen, J., Physical Review B, 105[11] 2022, 115202. https://doi.org/10.1103/PhysRevB.105.115202

[64] Anufriev, R., Nomura, M., Materials Today Physics, 15, 2020, 100272. https://doi.org/10.1016/j.mtphys.2020.100272

[65] Hu, Y., Li, D., Yin, Y., Li, S., Zhou, H., Zhang, G., RSC Advances, 10[42] 2020, 25305-25310. https://doi.org/10.1039/D0RA04737F

[66] Jenichen, A., Engler, C., Surface Science, 601[4] 2007, 900-907. https://doi.org/10.1016/j.susc.2006.11.043

[67] Liu, Z., Deng, F., Zhou, Y., Liang, Y., Peng, C., Peng, B., Zhao, F., Yang, Z., Chai, L., International Journal of Minerals, Metallurgy and Materials, 29[4] 2022, 662-670. https://doi.org/10.1007/s12613-022-2438-z

[68] Armington, A.F., Journal of Crystal Growth, 1[1] 1967, 47-48. https://doi.org/10.1016/0022-0248(67)90007-3

[69] Bouix, J., Hillel, R., Journal of Crystal Growth, 38[1] 1977, 61-66. https://doi.org/10.1016/0022-0248(77)90373-6

[70] Liu, Z., Deng, F., Zhou, Y., Liang, Y., Peng, C., Peng, B., Zhao, F., Yang, Z., Chai, L., International Journal of Minerals, Metallurgy and Materials, 29[4] 2022, 662-670. https://doi.org/10.1007/s12613-022-2438-z

[71] Xing, J., Chen, X., Zhou, Y., Culbertson, J.C., Freitas, J.A., Glaser, E.R., Zhou, J., Shi, L., Ni, N., Applied Physics Letters, 112[26] 2018, 261901. https://doi.org/10.1063/1.5038025

[72] Stukel, D.J., Physical Review B, 1[8] 1970, 3458-3463. https://doi.org/10.1103/PhysRevB.1.3458

[73] Bushick, K., Mengle, K., Sanders, N., Kioupakis, E., Applied Physics Letters, 114[2] 2019, 022101. https://doi.org/10.1063/1.5062845

[74] Prasad, C., Sahay, M., Physica Status Solidi B, 154[1] 1989, 201-207. https://doi.org/10.1002/pssb.2221540118

[75] Ku, S.M., Journal of the Electrochemical Society, 13, 1966, 813-816. https://doi.org/10.1515/9783110883558.bm

[76] Chu, T.L., Hyslop, A.E., Journal of Applied Physics, 43, 1972, 276. https://doi.org/10.1063/1.1661106

[77] Zaoui, A., El Haj Hassan, F., Journal of Physics Condensed Matter, 13[2] 2001, 253-262. https://doi.org/10.1088/0953-8984/13/2/303

[78] Boudjemline, A., Islam, M.M., Louail, L., Diawara, B., Physica B, 406[22] 2011, 4272-4277. https://doi.org/10.1016/j.physb.2011.08.043

[79] Nwigboji, I.H., Malozovsky, Y., Franklin, L., Bagayoko, D., Journal of Applied Physics, 120[14] 2016, 145701. https://doi.org/10.1063/1.4964421

[80] Buckeridge, J., Scanlon, D.O., Physical Review Materials, 3[5] 2019, 051601. https://doi.org/10.1103/PhysRevMaterials.3.051601

[81] Ku, S.M., Journal of the Electrochemical Society, 113[8] 1966, 813-816. https://doi.org/10.1149/1.2424125

[82] Chu, T.L., Hyslop, A.E., Journal of Applied Physics, 43[2] 1972, 276-279. https://doi.org/10.1063/1.1661106

[83] Chu, T.L., Hyslop, A.E., Journal of the Electrochemical Society, 121[3] 1974, 412-415. https://doi.org/10.1149/1.2401826

[84] Gujjar, D., Patel, P.C., Kandpal, H.C. Materials Today: Proceedings, 67, 2022, in press.

[85] Song, B., Chen, K., Bushick, K., Mengle, K.A., Tian, F., Gamage, G.A., Ren, Z., Kioupakis, E., Chen, G., Applied Physics Letters, 116[14] 2020, 141903. https://doi.org/10.1063/5.0004666

[86] Cohen, M.L., Materials Science and Engineering, 105-106[1] 1988, 11-18. https://doi.org/10.1016/0025-5416(88)90475-2

[87] Camacho-Mojica, D.C., López-Urías, F., Applied Surface Science, 368, 2016, 191-197. https://doi.org/10.1016/j.apsusc.2016.01.125

[88] He, Y., Sun, H., Physical Review Materials, 6[3], 2022, 034603.

[89] Brown, P.A., Shuford, K.L., Journal of Physical Chemistry C, 121[44] 2017, 24489-24494. https://doi.org/10.1021/acs.jpcc.7b06372

[90] Ullah, S., Denis, P.A., Sato, F., ACS Omega, 3[12] 2018, 16416-16423. https://doi.org/10.1021/acsomega.8b02605

[91] Zhou, Z.Z., Liu, H.J., Fan, D.D., Cao, G.H., Journal of Physics - Condensed Matter, 31[38] 2019, 385701. https://doi.org/10.1088/1361-648X/ab27f2

[92] Mishra, P., Singh, D., Sonvane, Y., Gupta, S.K., AIP Conference Proceedings, 2115, 2019, 030156.

[93] Hoat, D.M., Naseri, M., Hieu, N.N., Ponce-Pérez, R., Tong, H.D., Rivas-Silva, J.F., Vu, T.V., Cocoletzi, G.H., Superlattices and Microstructures, 139, 2020, 106399. https://doi.org/10.1016/j.spmi.2020.106399

[94] Da Ng, J., Danner, A., Physica Scripta, 96[5] 2021, 055801. https://doi.org/10.1088/1402-4896/abe0f1

[95] Zafari, M., Umer, M., Nissimagoudar, A.S., Anand, R., Ha, M., Umer, S., Lee, G., Kim, K.S., Journal of Physical Chemistry Letters, 13[20] 2022, 4530-4537. https://doi.org/10.1021/acs.jpclett.2c00918

[96] Xu, S., Song, B., Journal of Physical Chemistry C, 125[39] 2021, 21424-21433. https://doi.org/10.1021/acs.jpcc.1c05754

[97] Khossossi, N., Panda, P.K., Singh, D., Shukla, V., Mishra, Y.K., Essaoudi, I., Ainane, A., Ahuja, R., ACS Applied Energy Materials, 3[8] 2020, 7306-7317. https://doi.org/10.1021/acsaem.0c00492

[98] Yang, X., Tiwari, J., Feng, T., Materials Today Physics, 24, 2022, 100689.

https://doi.org/10.1016/j.mtphys.2022.100689

[99] Yu, C., Hu, Y., He, J., Lu, S., Li, D., Chen, J., Applied Physics Letters, 120[13] 2022, 132201. https://doi.org/10.1063/5.0086608

[100] Bushick, K., Chae, S., Deng, Z., Heron, J.T., Kioupakis, E., npj Computational Materials, 6[1] 2020, 3. https://doi.org/10.1038/s41524-019-0270-4

[101] Guan, Y., Li, X., Hu, Q., Zhao, D., Zhang, L., Applied Surface Science, 599, 2022, 153865. https://doi.org/10.1016/j.apsusc.2022.153865

[102] SanthiBhushan, B., Srivastava, A., Khan, M.S., Srivastava, A., Goumri-Said, S., IEEE Transactions on Electron Devices, 63[12] 2016, 4899-4906. https://doi.org/10.1109/TED.2016.2616387

[103] Dai, X., Zhang, X., Li, H., Applied Surface Science, 507, 2020, 144923. https://doi.org/10.1016/j.apsusc.2019.144923

[104] Guan, Y., Li, X., Hu, Q., Zhao, D., Zhang, L., Applied Surface Science, 599, 2022, 153865. https://doi.org/10.1016/j.apsusc.2022.153865

[105] Li, Y., Wei, D., Guo, G., Zhao, G., Tang, Y., Dai, X., Chinese Physics B, 31[9] 2022, 097301. https://doi.org/10.1088/1674-1056/ac6b2a

[106] Durandurdu, M., Journal of Non-Crystalline Solids, 524, 2019, 119656. https://doi.org/10.1016/j.jnoncrysol.2019.119656

[107] Daoud, S., Bioud, N., Lebga, N., Chinese Journal of Physics, 57, 2019, 165-178. https://doi.org/10.1016/j.cjph.2018.11.018

[108] Yaddanapudi, K., Computational Materials Science, 184, 2020, 109887. https://doi.org/10.1016/j.commatsci.2020.109887

[109] Davydov, S.Y., Semiconductors, 54[11] 2020, 1377-1380. https://doi.org/10.1134/S1063782620110068

[110] Li, S., Taddei, K.M., Wang, X., Wu, H., Neuefeind, J., Zackaria, D., Liu, X., Dela Cruz, C., Lv, B., Applied Physics Letters, 115[1] 2019, 011901. https://doi.org/10.1063/1.5103166

[111] Chen, X., Li, C., Tian, F., Gamage, G.A., Sullivan, S., Zhou, J., Broido, D., Ren, Z., Shi, L., Physical Review Applied, 11[6] 2019, 064070. https://doi.org/10.1103/PhysRevApplied.11.064070

[112] Bravić, I., Monserrat, B., Physical Review Materials, 3[6] 2019, 065402. https://doi.org/10.1103/PhysRevMaterials.3.065402

[113] Ge, Y., Wan, W., Guo, X., Liu, Y., Optics Express, 28[1] 2020, 238-248. https://doi.org/10.1364/OE.378374

[114] Tian, Z.Y., Zhang, Q.Y., Xiao, Y.W., Gamage, G.A., Tian, F., Yue, S., Hadjiev, V.G., Bao, J., Ren, Z., Liang, E., Zhao, J., Physical Review B, 105[17] 2022, 174306. https://doi.org/10.1103/PhysRevB.105.174306

[115] Wang, Y., Windl, W., Journal of Applied Physics, 129[7] 2021, 075703. https://doi.org/10.1063/5.0031005

[116] Chae, S., Mengle, K., Heron, J.T., Kioupakis, E., Applied Physics Letters, 113[21] 2018, 212101. https://doi.org/10.1063/1.5062267

[117] Viswanathan, E., Sundareswari, M., Journal of Chemical and Pharmaceutical Sciences, 11, 2015, 11-14.

[118] Tian, F., Luo, K., Xie, C., Liu, B., Liang, X., Wang, L., Gamage, G.A., Sun, H., Ziyaee, H., Sun, J., Zhao, Z., Xu, B., Gao, G., Zhou, X.F., Ren, Z., Applied Physics Letters, 114[13] 2019, 131903. https://doi.org/10.1063/1.5093289

[119] Kang, J.S., Li, M., Wu, H., Nguyen, H., Hu, Y., Applied Physics Letters, 115[12] 2019, 122103. https://doi.org/10.1063/1.5116025

[120] Mahat, S., Li, S., Wu, H., Koirala, P., Lv, B., Cahill, D.G., Physical Review Materials, 5[3] 2021, 033606. https://doi.org/10.1103/PhysRevMaterials.5.033606

[121] Malica, C., Dal Corso, A., Journal of Applied Physics, 127[24] 2020, 245103. https://doi.org/10.1063/5.0011111

[122] Malica, C., Dal Corso, A., Journal of Chemical Physics, 156[19] 2022, 194111. https://doi.org/10.1063/5.0093376

[123] Antonietti, M., Angewandte Chemie, 52[4] 2013, 1086-1087. https://doi.org/10.1002/anie.201207937

[124] Wang, S., Swingle, S.F., Ye, H., Fan, F.R.F., Cowley, A.H., Bard, A.J., Journal of the American Chemical Society, 134[27] 2012, 11056-11059. https://doi.org/10.1021/ja301765v

[125] Chen, Z., Zhao, J., Yin, L., Chen, Z., Journal of Materials Chemistry A, 7[21] 2019, 13284-13292. https://doi.org/10.1039/C9TA01410A

[126] Zhang, R.W., Zhang, C.W., Ji, W.X., Li, S.S., Wang, P.J., Hu, S.J., Yan, S.S., Applied Physics Express, 8[11] 2015, 113001. https://doi.org/10.7567/APEX.8.113001

[127] Broido, D.A., Lindsay, L., Reinecke, T.L., Physical Review B, 88[21] 2013, 214303. https://doi.org/10.1103/PhysRevB.88.214303

[128] Lindsay, L., Broido, D.A., Reinecke, T.L., Physical Review Letters, 111[2] 2013. 025901. https://doi.org/10.1103/PhysRevLett.111.025901

[129] Lv, B., Lan, Y., Wang, X., Zhang, Q., Hu, Y., Jacobson, A.J., Broido, D., Chen, G., Ren, Z., Chu, C.W., Applied Physics Letters, 106[7] 2015, 074105. https://doi.org/10.1063/1.4913441

[130] Ma, H., Li, C., Tang, S., Yan, J., Alatas, A., Lindsay, L., Sales, B.C., Tian, Z., Physical Review B, 94[22] 2016, 220303. https://doi.org/10.1103/PhysRevB.94.220303

[131] Hadjiev, V.G., Iliev, M.N., Lv, B., Ren, Z.F., Chu, C.W., Physical Review B, 89[2] 2014, 024308. https://doi.org/10.1103/PhysRevB.89.024308

[132] Chen, K., Song, B., Ravichandran, N.K., Zheng, Q., Chen, X., Lee, H., Sun, H., Li, S., Gamage, G.A.G.U., Tian, F., Ding, Z., Song, Q., Rai, A., Wu, H., Koirala, P., Schmidt, A.J., Watanabe, K., Lv, B., Ren, Z., Shi, L., Cahill, D.G., Taniguchi, T., Broido, D., Chen, G., Science, 367[6477] 2020, 555-559. https://doi.org/10.1126/science.aaz6149

[133] Rai, A., Li, S., Wu, H., Lv, B., Cahill, D.G., Physical Review Materials, 5[1] 2021, 013603. https://doi.org/10.1103/PhysRevMaterials.5.013603

[134] Kim, J., Evans, D.A., Sellan, D.P., Williams, O.M., Ou, E., Cowley, A.H., Shi, L., Applied Physics Letters, 108[20] 2016, 201905. https://doi.org/10.1063/1.4950970

[135] Kang, J.S., Li, M., Wu, H., Nguyen, H., Hu, Y., Science, 361[6402] 2018, 575-578. https://doi.org/10.1126/science.aat5522

[136] Tian, F., Song, B., Lv, B., Sun, J., Huyan, S., Wu, Q., Mao, J., Ni, Y., Ding, Z., Huberman, S., Liu, T.H., Chen, G., Chen, S., Chu, C.W., Ren, Z., Applied Physics Letters, 112[3] 2018, 031903. https://doi.org/10.1063/1.5004200

[137] Mei, H., Xia, Y., Zhang, Y., Wu, Y., Chen, Y., Ma, C., Kong, M., Peng, L., Zhu, H., Zhang, H., Physical Chemistry Chemical Physics, 24[16] 2022, 9384-9393. https://doi.org/10.1039/D1CP05979C

[138] Lyons, J.L., Varley, J.B., Glaser, E.R., Freitas, J.A., Jr., Culbertson, J.C., Tian, F., Gamage, G.A., Sun, H., Ziyaee, H., Ren, Z., Applied Physics Letters, 113[25] 2018, 251902. https://doi.org/10.1063/1.5058134

[139] Protik, N.H., Carrete, J., Katcho, N.A., Mingo, N., Broido, D., Physical Review B,

94[4] 2016, 045207. https://doi.org/10.1103/PhysRevB.94.045207

[140] Meng, X., Singh, A., Juneja, R., Zhang, Y., Tian, F., Ren, Z., Singh, A.K., Shi, L., Lin, J.F., Wang, Y., Advanced Materials, 32[45] 2020, 2001942. https://doi.org/10.1002/adma.202001942

[141] Zhou, Y., Hsieh, W,P., Chen, C,C., Meng, X., Tian, F., Ren, Z., Shi, L., Lin, J,F., Wang, Y., Applied Physics Letters, 121[12] 2022, 121902. https://doi.org/10.1063/5.0113007

[142] Ravichandran, N.K., Broido, D., Nature Communications, 10[1] 2019, 827. https://doi.org/10.1038/s41467-019-08713-0

[143] Hou, S., Sun, B., Tian, F., Cai, Q., Xu, Y., Wang, S., Chen, X., Ren, Z., Li, C., Wilson, R.B., Advanced Electronic Materials, 89[10] 2022, 2200017. https://doi.org/10.1002/aelm.202200017

[144] Zhou, Y., Hsieh, W.P., Chen, C.C., Meng, X., Tian, F., Ren, Z., Shi, L., Lin, J.F., Wang, Y., Applied Physics Letters, 121[12] 2022, 121902. https://doi.org/10.1063/5.0113007

[145] Ouyang, Y., Yu, C., He, J., Jiang, P., Ren, W., Chen, J., Physical Review B, 105[11] 2022, 115202. https://doi.org/10.1103/PhysRevB.105.115202

[146] Chen, X., Li, C., Xu, Y., Dolocan, A., Seward, G., Van Roekeghem, A., Tian, F., Xing, J., Guo, S., Ni, N., Ren, Z., Zhou, J., Mingo, N., Broido, D., Shi, L., Chemistry of Materials, 33[17] 2021, 6974-6982. https://doi.org/10.1021/acs.chemmater.1c02006

[147] Fava, M., Protik, N.H., Li, C., Ravichandran, N.K., Carrete, J., van Roekeghem, A., Madsen, G.K.H., Mingo, N., Broido, D., npj Computational Materials, 7[1] 2021, 54. https://doi.org/10.1038/s41524-021-00519-3

[148] Magoulas, I., Kalemos, A., Journal of Chemical Physics, 139[15] 2013, 154309. https://doi.org/10.1063/1.4824886

[149] Lee, H., Gamage, G.A., Lyons, J.L., Tian, F., Smith, B., Glaser, E.R., Ren, Z., Shi, L., Journal of Physics D, 54[31] 2021, 31LT01. https://doi.org/10.1088/1361-6463/abfefa

[150] Liu, Z., Yang, X., Zhang, B., Li, W., ACS Applied Materials and Interfaces, 13[45] 2021, 53409-53415. https://doi.org/10.1021/acsami.1c11595

[151] Dutta, M., Prasad, M.V.D., Pandey, J., Soni, A., Waghmare, U.V., Biswas, K., Angewandte Chemie 61[15] 2022, 202200071.

https://doi.org/10.1002/anie.202200071

[152] Pan, F., Gamage, G.A., Sun, H., Ren, Z., Journal of Applied Physics, 131[5] 2022, 055102. https://doi.org/10.1063/5.0073394

[153] Shi, H.X., Yang, K.K., Luo, J.W., Acta Physica Sinica, 70[14] 2021, 147302. https://doi.org/10.7498/aps.70.20210797

[154] Sun, H., Chen, K., Gamage, G.A., Ziyaee, H., Wang, F., Wang, Y., Hadjiev, V.G., Tian, F., Chen, G., Ren, Z., Materials Today Physics, 11, 2019, 100169. https://doi.org/10.1016/j.mtphys.2019.100169

[155] Zhou, Y., Li, C., Koirala, P., Gamage, G.A., Wu, H., Li, S., Ravichandran, N.K., Lee, H., Dolocan, A., Lv, B., Broido, D., Ren, Z., Shi, L., Physical Review Materials, 6[6] 2022, L061601. https://doi.org/10.1103/PhysRevMaterials.6.L061601

[156] Yue, S., Gamage, G.A., Mohebinia, M., Mayerich, D., Talari, V., Deng, Y., Tian, F., Dai, S.Y., Sun, H., Hadjiev, V.G., Zhang, W., Feng, G., Hu, J., Liu, D., Wang, Z., Ren, Z., Bao, J., Materials Today Physics, 13, 2020, 100194. https://doi.org/10.1016/j.mtphys.2020.100194

[157] Wu, H., Fan, H., Hu, Y., Physical Review B, 103[4] 2021, L041203. https://doi.org/10.1103/PhysRevB.103.L041203

[158] Ziyaee, H., Gamage, G.A., Sun, H., Tian, F., Ren, Z., Journal of Applied Physics, 126[15] 2019, 155108. https://doi.org/10.1063/1.5110890

[159] Gamage, G.A., Sun, H., Ziyaee, H., Tian, F., Ren, Z., Applied Physics Letters, 115[9] 2019, 092103. https://doi.org/10.1063/1.5111732

[160] Guo, J., Xu, G., Tian, D., Qu, Z., Qiu, C.W., Advanced Materials, 34[17] 2022, 2200329. https://doi.org/10.1002/adma.202200329

[161] Kang, J.S., Li, M., Wu, H., Nguyen, H., Aoki, T., Hu, Y., Nature Electronics, 4[6] 2021, 416-423. https://doi.org/10.1038/s41928-021-00595-9

[162] Tavakoli, A., Vafai, K., Journal of Heat Transfer, 143[10] 2021, 102901. https://doi.org/10.1115/1.4050922

[163] Tavakoli, A., Salimpour, M.R., Vafai, K., Numerical Heat Transfer A, 80[8] 2021, 389-410. https://doi.org/10.1080/10407782.2021.1947626

[164] Wei, Z., Yang, Z., Liu, M., Wu, H., Chen, Y., Yang, F., Journal of Applied Physics, 127[5] 2020, 055105. https://doi.org/10.1063/1.5139669

www.ingramcontent.com/pod-product-compliance
Lightning Source LLC
Chambersburg PA
CBHW071713210326
41597CB00017B/2465

*9 7 8 1 6 4 4 9 0 2 2 2 6 *